CAMBRIDGE SOCIAL
BIOLOGY TOPICS

Series editors
S. Tyrell Smith and Alan Cornwell

BEHAVIOUR AND SOCIAL ORGANISATION

Michael Reiss and Harriet Sants
Hills Road Sixth Form College, Cambridge
MRC Unit on the Development and Integration of Behaviour, Cambridge

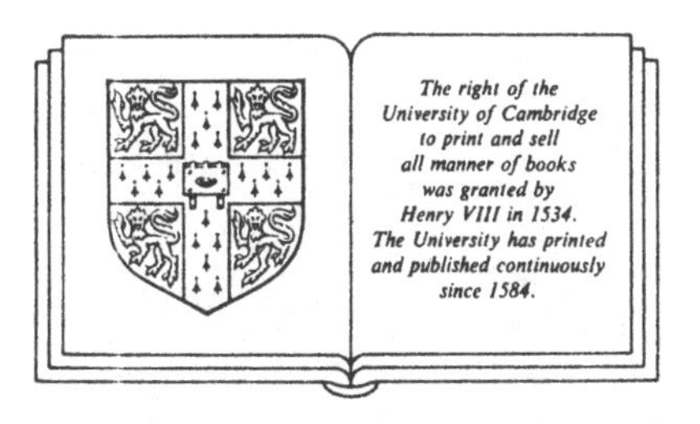

CAMBRIDGE UNIVERSITY PRESS

Cambridge
New York New Rochelle
Melbourne Sydney

CAMBRIDGE UNIVERSITY PRESS
Cambridge, New York, Melbourne, Madrid, Cape Town, Singapore, São Paulo, Delhi

Cambridge University Press
The Edinburgh Building, Cambridge CB2 8RU, UK

Published in the United States of America by Cambridge University Press, New York

www.cambridge.org
Information on this title: www.cambridge.org/9780521313285

First published 1987
Second printing 1988
Re-issued in this digitally printed version 2009

A catalogue record for this publication is available from the British Library

ISBN 978-0-521-31328-5 paperback

Illustrations by
Jake Tebbitt and Marlborough Design
Cover photo by Harriet Sants
Cover photo shows children at play

Contents

Authors' acknowledgements

We should like to thank the following for their invaluable help and encouragement: Pat Bateson, Jenny Chapman, Alan Cornwell, Judy Dunn, Paul Martin, John Sants and Stephen Tomkins.

Acknowledgements

The publishers would like to thank the following for permission to reproduce copyright material:
2.2, *Journal of Biological Education* **18**, 3 (Reiss 1984); drawings by Dr Jenny L. Chapman. 2.4, Harvard University Press © 1975, redrawn from E. O. Wilson *Sociobiology*. 3.1, Harcourt Brace Jovanovich © 1979, redrawn from Hilgard *et al. Introduction to psychology*, seventh edition; Table 8.1 © 1980, from Maccoby *Social development*. 3.2, 3.4, Blackwell Scientific Publications, redrawn from Halliday *et al.* (1983) *Animal behaviour, three genes development and learning*; 4.2, 4.3, 4.5, from Krebs *et al.* (1978, 1984) *Behavioural ecology*. 4.4, Dr Brian Bertram. 4.6, Sinauer Associates © 1975, redrawn from Alcock *Animal behaviour*, first edition. 4.7, Scientific American © 1971, redrawn from Watts *et al. The social order of turkeys*. 5.1, Chicago University Press © 1982, redrawn from Clutton-Brock *et al. Red deer: the behaviour and ecology of two sexes.* 6.1, H. F. and G. Witherby, redrawn from Lack (1965) *The life of the robin*. 6.1, Dr Tim Clutton-Brock. Tables 7.1, 7.2, 7.4, Macmillan, from Nobbs (1983) *Sociology in context*. 8.1, University of Wisconsin Private Laboratory. 8.2, Harriet Sants.

1 Introduction

This book is concerned with the organisation of society and the development of social behaviour in man and other animals. The areas of study it covers have been given a wide variety of names. The approaches used are numerous. In this introduction we shall attempt to outline some of these approaches and to point to some of the important questions and issues that will arise.

The study of any kind of behaviour is really an attempt to answer the question 'Why do animals or people behave in the way that they do?' Niko Tinbergen, a very well-known student of animal behaviour, identified four different kinds of 'why' questions:

1 What are the mechanisms underlying it?
2 How did it develop?
3 What is it for?
4 How did it evolve?

For example, the question 'Why does the thumb move in a different way from the other fingers?' can be answered in four different ways. Firstly, it can be answered in terms of how muscle and bone move; secondly, in terms of the hands' embryology, in other words how they develop in the womb; thirdly, one might say that it makes it easier to pick things up; and fourthly, that humans have descended from monkey-like creatures and that these had opposable thumbs, so humans do too. This book is mainly concerned with tackling the third question. It also deals with the second and fourth questions but hardly at all with the first. As we shall see, this means that 'why' questions are answered in terms of function, evolution and development.

The studies and ideas described have been drawn from a wide range of disciplines including ethology, behavioural ecology, psychology, sociobiology and sociology. However, the fact that these names exist does not necessarily mean that the areas of study are distinct. All scientists in these fields are interested in behaviour but they vary in terms of their approaches, methods and major questions. Behavioural ecologists are interested in how particular behaviour may contribute to the survival of an animal. For example, they may try to answer the question 'How does living in a group contribute to an animal's survival?' Ethologists tend to approach the study of animal behaviour from clear descriptions of the behaviour, particularly by watching the animals in the wild. Where many non-ethological biologists have concentrated on the immediate causes and development of behaviour, ethologists are also interested in answering the questions 'What is it for and how did it evolve?' (questions 3 and 4 of Tinbergen's four 'whys').

Sociobiology may be defined as the study of the biological basis of social behaviour, specifically in terms of its function and evolution. Although scientists had been tackling questions about social behaviour from a biological point of view for many years before, sociobiology was in a sense christened by the publication in 1975 of a major book by E. O. Wilson called *Sociobiology: the new synthesis*. This aimed to bring together ethology and animal behaviour on the one hand and the study of genetics and evolution on the other. Although there have been some who have argued that most if not all social behaviour, even of humans, can be described and explained from a biological perspective, this has as yet not always been accepted. Many useful ideas and explanations have also been produced by sociologists and psychologists. Therefore, in order to get a broader perspective on social behaviour, this book tries to give the reader tastes of a whole range of perspectives and approaches, not just the biological. All are valid and none can really substitute for another. Their usefulness depends on the kind of question that is required to be answered and the particular interest of the investigator.

Many early writers on behaviour were concerned with sorting out what aspects of behaviour are brought about by instinct (sometimes referred to as innate behaviour), and what by learning. With regard to human behaviour, the concern was also over which behaviour could be described as 'biological' and which as cultural. Nowadays, these questions are usually regarded as being inappropriate. Most behaviour is seen as being not entirely instinctive (i.e. developing completely independently of any environmental influences) but as arising as a result of an interaction between an animal's genetic predispositions and the environment, which often involves learning. However, the term innate has come to be used by some to describe activities characteristic of the species. By this definition innate behaviour describes behaviour which tends not to vary greatly between individuals of the same species, though it may involve learning or be otherwise influenced by the environment.

This book is concerned with the social behaviour of both non-human species of animals and humans. One important question to consider while reading it is how far the study of non-human species is relevant and can help in the understanding of the behaviour of man. Man is an animal and as such will have been and is subjected to evolutionary pressures and biological constraints, as are other animals. On the other hand, all species of animals have their own particular characteristics adapted to particular environments. Man differs from other species in some very important ways. Man possesses an enormous potential for learning and a highly complex language. This has enabled him to develop varied and complex cultures with a wide range of beliefs and norms. The methods and frameworks used by biologists may not always be appropriate for studying these aspects of man's society. This is why ideas and research drawn from the non-biological sciences are also presented in this book. A comparison between man and other species can be useful but it should be made with caution, as indeed should comparisons between any species. It may be possible, as we shall see, to make up a story entirely based on, for example, the function or significance for survival for early man of a piece of human behaviour, but this may be so far removed from its present form that it helps very little in understanding why and under what circumstances that particular behaviour occurs now. One good example of this is war. Aspects of aggression between individuals in man may well be similar to aggression in other animals but the occurrence of modern warfare is unlikely to be explained without reference to historical, political and economic factors, all of which are peculiar to man's culture. A complete explanation of man's social behaviour must be drawn from disciplines that at the moment are given different names, for example sociology, psychology and sociobiology.

2 The functions and evolution of behaviour

When we see a cat crouched on the ground, motionless except for the occasional swish of its tail, staring at a blackbird feeding a few metres away, we have little hesitation in suggesting functions of the cat's behaviour: 'It's stalking its prey' or 'It's trying not to be seen'. Suppose, now, that our domestic predator rushes unsuccessfully at its intended prey which disappears squawking into a nearby tree. What could we suggest as the function of the blackbird's cry? Does it serve to warn other blackbirds? Is it a desperate attempt to startle the cat? Or does the cry merely release tension? Ethologists from before the time of Charles Darwin have suggested explanations for the behaviour they observed in the natural world. Only more recently, however, have experiments been performed to test such hypotheses. Such experiments, and the critical thinking they typically engender, have helped to raise the study of animal behaviour to a science.

2.1 Black-headed gulls

During the 1950s and 1960s, the Dutch ethologist Niko Tinbergen, based at Oxford University, led a large research programme into the behaviour of gulls – particularly black-headed gulls, herring gulls and kittiwakes. The work on black-headed gulls was done in Cumbria.

Life in the colony

Black-headed gulls, *Larus ridibundus*, are social, crowding their nests together even when other apparently suitable nest sites are available elsewhere. Within a colony there is a system of **territories**. These are the properties of individual pairs of gulls, so that a male and a female between them defend their territory. The birds are normally **monogamous**, so that one female pairs with one male. Some birds arrive paired, others form pairs after arrival at the colony. Both partners incubate the eggs and subsequently protect and feed the young.

Early in spring the birds gradually return to the colony from their winter quarters. As soon as the birds arrive on the breeding grounds, much calling and posturing are evident. At first, Tinbergen and his co-workers had little idea what all the commotion meant. But when they watched individual birds closely, they saw that the birds spent their whole time intruding into each others' territories, withdrawing when the owner called or postured, and rushing up to strangers when these in turn trespassed. When other birds intrude, the pair, particularly the male, respond by either attacking them outright or by posturing and calling. Actual fights are rare. When they do occur, fighting is done by delivering vigorous pecks with the strong bill, usually from above or, when two birds come to grips, by striking the rival with half-folded wings.

Eggs and eggshells

After all the courtship and initial territorial defence the eggs are laid, incubated and eventually hatched. Then, within a few hours after hatching, an intriguing piece of behaviour occurs. One of the parents takes the empty shell in its bill, walks or flies away with it, and drops it well away from the nest. It took Tinbergen a long time to become interested in this, because it seemed such a minor piece of behaviour. However, as Tinbergen began to see more of predation, he noticed that predators, such as neighbouring black-headed gulls, marauding herring gulls or passing carrion crows, were often on the alert for just such occasions when a gull leaves its brood unprotected for a moment. Removing the eggshell exposes the brood to a real risk.

Tinbergen had several hypotheses for the function of eggshell removal. First, their sharp edges might occasionally injure the delicate chick. This has been reported sometimes to be the case in duck hatcheries. Secondly, an empty shell might slip over an as-yet-unhatched egg, and imprison the chick inside. Thirdly, a gull has only three brood spots – defeathered patches on its belly – each of which the incubating bird brings carefully into contact with one egg, and empty shells in the nest might 'compete' for a brood spot with an egg or a just-hatched chick. A fourth possibility is that the moist organic waste that always remains behind in the shell might be a favourable incubation ground for disease-causing bacteria. Finally, the eggshell, by showing its white rims and inside, might attract the attention of predators, in particular those such as foxes that walk through a colony during the half-dark summer nights, or 'dart-and-grab' raiders such as carrion crow that have to spot their quarry in a flock.

Tinbergen was particularly attracted to this last hypothesis because a related species, the kittiwake, does not bother to remove its eggshell, or does so in a very lackadaisical manner. The kittiwake is a cliff-nesting species where predation is much less of a problem. It is noteworthy that kittiwake chicks are unique among gulls in that they are not camouflaged; instead of the buff ground colour and a pattern of irregular dark dots, usual in the downy plumage of gulls, they have a beautiful silvery sheen.

To test whether the removal of eggshells might reduce the risk of predation in black-headed gulls, Tinbergen laid out a number of well-scattered, isolated eggs in a valley adjoining the gulleries. To half of these were added, at a distance of five centimetres, an empty eggshell. The remaining half had no such neighbours. Tinbergen and his associates then retired to a hide on a nearby dune-top. Carrion crows, herring gulls and an occasional black-headed gull swooped down and attacked intact eggs. When it was estimated that about half the eggs had been taken, the results were determined. Many more 'lone' eggs survived than did eggs which had an eggshell nearby. The tests were repeated with different distances between the eggs and the empty shells (see Figure 2.1). The results were clear. The greater the distance, the lower the predation. Observation of predatory birds showed what happened. The crow or

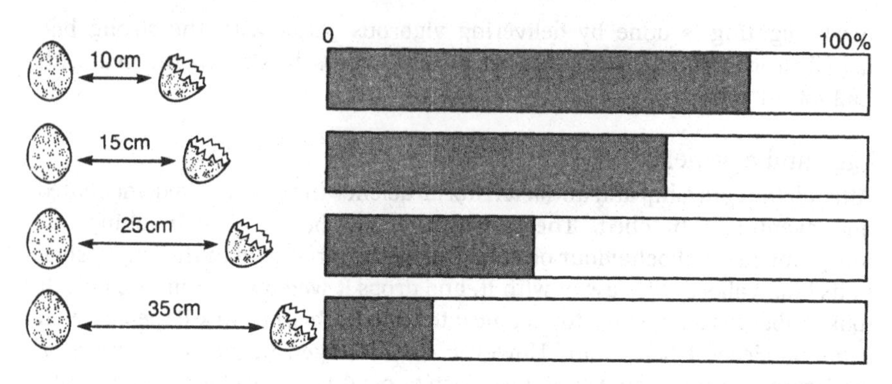

Figure 2.1 The further away an eggshell is from an unhatched egg, the less chance there is of the egg being discovered by a predator. The lengths of the grey bars show the percentages of eggs taken, within a standard distance, for each of the egg-to-eggshell distances indicated.

gull would alight near to, or walk up to, the empty shell and then start to walk around it. An intact egg at a distance of five centimetres was almost invariably found, despite its natural speckled camouflage. Eggs further away were often overlooked, and the bird lost interest. So here is at least one advantage of eggshell removal – it reduces the risk of predation.

2.2 Sticklebacks

Sticklebacks are small freshwater fish. In a series of experiments dating from the 1930s, Tinbergen investigated the reproductive behaviour of the three-spined stickleback, *Gasterosteus aculeatus*.

The classic stickleback account

In spring, male sticklebacks set up territories from which they chase away intruders of either sex. During this period males acquire red bellies. A patch of red acts as a **sign stimulus**, that is, it provokes a stereotyped response. The red patch makes males particularly likely to be chased away. By experimenting, Tinbergen found that a realistically shaped but non-red model male stickleback provoked little interest from a territorial male, whereas extremely crude models painted red on their lower surfaces provoked strong aggression. His nest complete, the male now becomes interested in females swollen with eggs. When a female appears he moves towards her in a curious zig-zag fashion (the zig-zag dance). When she sees him, the female responds by swimming towards the male with her head and tail turned upwards, thereby displaying her swollen abdomen. The swollen abdomen, too, acts as a sign stimulus: realistic model females lacking a swollen belly are not courted, while crude females provided with a swollen lower surface are. Having displayed her swollen abdomen to the male, the female is then led by the male, who shows her the nest entrance by poking it with his snout. She enters the nest and the male gives the female's rump several prods with a trembling motion, and this stimulates her to lay the eggs. When she has discharged all her eggs, she leaves the nest and the male enters it and ejaculates sperm over the eggs. He then

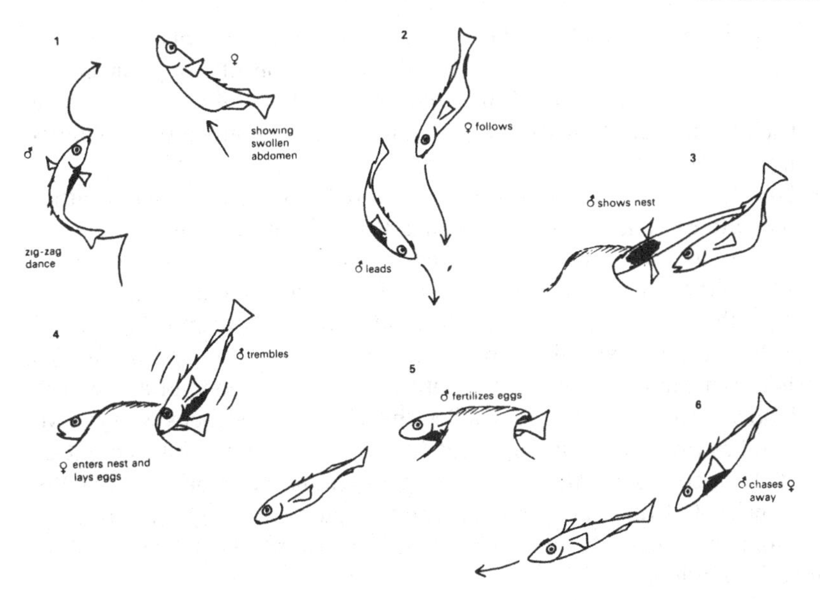

Figure 2.2 Courtship and reproduction in the three-spined stickleback.

chases the female away (see Figure 2.2). A male may mate with as many as five different females; then, having fertilised several clutches, he loses his readiness to court females. Instead he begins regular ventilation of the eggs by fanning them with his pectoral fins. The time spent in fanning increases daily until the eggs hatch, when it stops. Fanning appears to function as a means of aerating the eggs. Evidence for this is provided by the observation that artificial lowering of the oxygen content of the water induces increased fanning.

This elegant story has remained essentially unchanged for over 30 years. Recently, however, three separate studies have been performed. Taken together these constitute a major amendment to the usual description of stickleback courtship and reproduction.

Recent stickleback research

Perhaps the most remarkable alteration to the classic account is provided by Rowland. Rowland carefully repeated Ter Pelkwijk and Tinbergen's experiments on the importance of stickleback coloration in territorial encounters between males. His results are easily summarised: adding red to a model stickleback male makes it less likely to be attacked by a territorial male! This conclusion is the exact opposite of the one originally reported by Ter Pelkwijk and Tinbergen and now a part of the standard account. Such an emphatic contradiction of perhaps the most famous of all ethological experiments seems remarkable, but was statistically significant.

Therein, perhaps, lies an important point. Rowland carefully **replicated** his experiments and tested his hypothesis using accepted **statistical tests.** Tinbergen merely, one might say anecdotally, reported his 'finding'. It may be unfair to criticise Tinbergen with the benefit of hindsight. Certainly Tinbergen wasn't alone in eschewing statistics. Konrad Lorenz, Tinbergen's close friend and colleague, with whom he and Karl von Frisch shared the only Nobel Prize ever awarded for ethology, openly boasted that none of his publications contained any graphs or tables! Times have changed, and today the average ethologist has a familiarity with statistics that would put many other biologists to shame.

Rowland also pointed out that the fact that red coloration prevents rather than elicits attack is as might be expected. It is difficult to imagine how red coloration could have evolved if its effect was to release aggression in opponents. This mode of thinking is typical of modern ethological thought. In postulating the way in which a 'behaviour' (here the acquisition of red coloration) may have evolved, it is necessary to consider as many implications as possible that the new behaviour might have for the fitness of its owner.

Rowland's results directly contradict the classic stickleback story; two further studies (by Li and Owings, and Ridley and Rechten) supplement or modify it. They considered the behaviour of the female sticklebacks, while the earlier work of Tinbergen's school concentrated on the male role in reproduction to the extent that females were viewed as the passive recipients of the males' attentions.

A crucial factor in the research carried out by Li and Owings was that, unlike previous researchers, they watched female sticklebacks for as long as they watched males. Their observations contrast with the popular notion that female sticklebacks are relatively passive participants in reproduction. Before individuals of either sex were introduced to individuals of the other sex, they were sexually segregated into groups of six. This enabled Li and Owings to determine female–female interactions as well as male–male interactions. There were more aggressive encounters per hour in the all-female groups than in the all-male groups. Some females defended territories, others did not. What is the function of this female aggression? On three occasions (two of which were observed) females actually lost their eggs. Each time this was observed, a dominant female poked or squashed a subordinate female, and repeated attacks led to the two subordinate females prematurely shedding their eggs. Li and Owings also studied groups consisting of six females and six males. Female–female interactions were again important. On three occasions a subordinate female accepted a courting male before the dominant. Each time the dominant female disrupted courtship. On the two occasions that a subordinate female attempted to disrupt courtship by a dominant female, the attempts failed.

The third relevant study on stickleback reproductive behaviour was carried out by Ridley and Rechten. It was known to Tinbergen that females sometimes refuse to spawn with courting males, but before Ridley and Rechten's study, female refusal had received little attention. Ridley and Rechten found that the presence of eggs in a male's nest increased the chance that a female would spawn with him. Their findings suggest that female refusal has a function. The hypotheses suggested by Ridley and Rechten, and the evidence derived from experimental observation and from theoretical argument which they marshal for and against these hypotheses, illustrate current ethological methodology.

Ridley and Rechten suggest three reasons why female sticklebacks prefer to spawn with males whose nests contain eggs. First, male sticklebacks might occasionally eat some of their offspring. The more eggs there are, the lower the chance that a particular egg will be eaten. Secondly, the same argument applies to other predators. Finally, it seems likely that the risk of predation on the eggs is especially high during the courtship phase, as only during this phase does the male allow other fish near his nest. The time from the laying of the eggs to the beginning of the parental phase is shorter for later clutches. The risks of predation should, therefore, be lower for later clutches. By laying in a nest that already contains eggs, a female ensures that she is laying one of the later clutches.

In Ridley and Rechten's experiments, males with eggs were no more likely to be followed by females than were males without eggs. Females are apparently unable to distinguish between males with and without eggs. This makes functional sense because males that **signalled** in any way that they had no eggs would be selected against by females.

Ridley and Rechten point out that their results suggest new functional explanations for two other components of stickleback courtship. First, **egg kidnapping.** Territorial male sticklebacks sometimes kidnap eggs from neighbouring nests and rear them. Once females prefer to mate with males who already have eggs in their nests, egg kidnapping by males will be adaptive. Secondly, **displacement fanning.** During stickleback courtship so-called displacement fanning often occurs. The male temporarily leaves the female, goes to the nest and makes those movements by which males normally fan the eggs, so supplying them with the necessary oxygen. Such displacement fanning may occur whether or not the male actually has any eggs in the nest. Fanning by males during stickleback courtship has been considered the perfect example of a functionless **displacement activity**. Ridley and Rechten suggest the following evolutionary sequence: originally only males with eggs fanned; females could then have chosen males with eggs by only following those that fanned. Males without eggs would then be selected, by natural selection, to fan when a female appeared.

2.3 Supernormal sign stimuli

Supernormal sign stimuli were first discovered in a study of egg recognition in the ringed plover. If presented with a normal egg (which is light brownish with darker brown spots) and an egg with a clear white ground and black dots, the birds preferred the latter type. A more dramatic example of a supernormal stimulus is afforded by oystercatchers in work again carried out by Tinbergen. If an oystercatcher is presented with an egg of normal oystercatcher size, one of herring gull's size and one double the (linear) size of a herring gull's egg, the majority of choices fall upon the largest egg, despite the fact that even a normal herring gull's egg is much larger than an oystercatcher's egg (Figure 2.3). The explanation for such abnormal behaviour is presumably that in real life, the larger an egg, the greater its survival chances and potential reproductive success.

Figure 2.3 An oystercatcher attempting to incubate a supernormal egg in preference to a herring gull's egg (right) and a normal oystercatcher egg (middle).

Another example of supernormal stimuli is afforded by courtship in the silver-washed fritillary butterfly, *Argynnis paphia*. The normal wing area of a female silver-washed fritillary is about 20 cm^2, and in flight she beats her wings about 7 or 8 times per second. In experiments, however, where males were given the choice between model females of different sizes, the preferred female turns out to have a wing area of 88 cm^2. Similarly, models that flick their wings at the extraordinarily high rate of 70 beats per second are chosen by males. Interestingly, investigations reveal that an area of 88 cm takes up the whole of the male's visual field, at the distance the models were presented, while 70 beats per second equals the **flicker-fusion frequency** for silver-washed fritillaries. The flicker-fusion frequency is the point at which separate objects presented rapidly in succession appear to fuse together. In humans, it equals about 20 objects per second, which is why images presented at a rate greater than this – as, for example, in the separate frames that together comprise a movie – cannot be distinguished, but rather fuse together. It appears, therefore, that male silver-washed fritillaries simply want their mates to be as large as possible, and to flap their wings as fast as possible. What functions, if any, do these preferences serve?

It is very well established in butterflies and most other species that larger females produce more offspring. So males who prefer large mates will father more offspring. But why should males like their mates to flap their wings as fast as possible? The answer isn't known, but we can suggest several possible explanations. Maybe such females are demonstrating that they are healthy and capable of avoiding predators or feeding more. Maybe such females can fly further and so disperse eggs over a wider area. Or perhaps the male's behaviour is functionless – an inevitable response determined by the physiology of the insect eye.

2.4 The evolution of behaviour

Already, in our consideration of the function of behaviour, we have found ourselves hypothesising on the evolutionary history of behaviour. This is typical of ethology. A full understanding of the function of a behaviour only comes from an understanding of how the behaviour evolved, the mechanisms by which it operates and the way in which it develops in the life of each individual (see Chapter 1).

Behaviour does not fossilise in the way that bones or pollen grains do. It might be thought that the fossil record, therefore, tells us nothing about the evolution of behaviour. This is something of an oversimplification. For example, the recent discovery of fossilised dinosaur nests immediately reveals something of the behaviour of both parents and young. Similarly, there is a whole science of **trace fossils**, investigating the traces animals leave behind as a result of their movements or homes they constructed. Nevertheless, it is true that the study of the evolution of behaviour relies heavily on the **comparative method**.

2.5 The comparative method

The comparative method relies on examining the behaviour of present-day species and trying to construct plausible evolutionary sequences. It is similar, therefore, to the study of comparative anatomy, and indeed its originator, Konrad Lorenz, maintained that a piece of behaviour could be considered just as an anatomist would consider a bone in a skeleton.

An elegant example of the use of the comparative method is in the study of how an unusual courtship signal evolved in a small fly, *Hilara sartor*. The male of this species constructs a delicate hollow silk balloon that is almost as large as he is. He then flies to a swarm composed of other balloon-carrying males, there to circle about until a female arrives (Figure 2.4). She selects a mate from the swarm, accepts his balloon, and the two leave the other males and copulate.

Figure 2.4 Courtship in the empid fly, *Hilara sartor*. Four displaying males are shown. A female (♀) is in the act of accepting an empty silk balloon from one of the males as a precondition for mating.

How did this behaviour evolve? Without comparative data the puzzle might remain unsolved. Fortunately there are thousands of species of empid flies (the family to which *H. sartor* belongs). E. L. Kessel has summarised the available information and divides empids into eight groups on the basis of their reproductive behaviour.

1. Flies in this group, and all others, except for those in groups 7 and 8, are at least partly carnivorous, hunting for other small flies. Males belonging to group 1 search for a female and court her in isolation from others.
2. Males in this group capture prey before locating a female. She takes the food prior to copulation and consumes it during mating.
3. Males capture prey and then form swarms. A female attracted to the swarm selects a male, receives the prey, and mates.
4. The behaviour of group 4 flies is very similar to that of group 3 flies but here the male applies some strands of silk to the prey prior to joining the swarm.

5 The same as in group 4, except that the male wraps the prey entirely in a heavy silk bandage before offering it to a female.
6 The same as in group 5, except that the male removes the juices from his offering prior to wrapping it. As a result the female receives a non-nutritious husk.
7 These species feed only on nectar. However, prior to courtship the male finds a dried insect fragment and uses it as a foundation for the construction of a large balloon, which he then presents to a female before copulating.
8 *Hilara sartor* and a few other species omit an insect fragment as the starting point for balloon construction.

One can speculate about the impetus for each evolutionary change. Kessel and others have suggested that the initial 'wedding gift' invitation is advantageous because of the predatory nature of the female fly and consequent risk to the male of cannibalism by the female. The present can be viewed as a distraction for the female, so that the male can get on with reproduction – safely. An alternative and not mutually exclusive hypothesis is that the gift of prey gives the female a high-protein meal that helps her eggs to mature rapidly. Thus it is to her advantage to choose a present-giving male if one is available, and it may be further to the male's advantage as well, if the prey assists in the development and survival of eggs he has fertilised. The continued existence of group 1 courtships in empids could stem simply from the failure of any variant male in these species to offer a female a fly while courting. Or there may be something about their ecology that reduces the risk of cannibalism or lowers the advantage to a male of giving the female an energy boost.

The same kind of speculation can be applied to every other change. Collecting in swarms may make males more conspicuous to females. Once some males swarm, females should avoid solitary males as it is to a female's advantage to have a chance to compare males (in a swarm) and to choose to copulate with the male carrying the best present (group 3). The silk may serve to retain prey initially (group 4) and then to make it look larger than it really is (group 5). By this stage the balloon has become the releaser of mate choice for the female, and the prey itself is no longer critical. A male can 'cheat' the female and remove the juices of the prey (group 6). Upon adoption of a purely nectivorous feeding niche (group 7), capture of prey serves no nutritional function, yet continues. Maybe the male requires the insect as a releaser of balloon building, or maybe the female requires the insect as a releaser of mating behaviour. Finally, this component is lost (group 8).

2.6 Behavioural genetics

If behaviour is to evolve by natural selection, then behavioural differences between individuals need to have a genetic basis. Consideration of the breeds of domestic dog reveals how much behaviour must have been inherited – docility, aggression, stealth, a readiness to bark, mouthing to kill and mouthing to retrieve, for example.

One of the most famous examples of the genetic basis of behaviour involves the nest-clearing behaviour of honeybee workers. Bees in some hives have the ability to perform two behaviour patterns: first, uncapping cells that contain diseased pupae; and second, removing the dead pupae from the cells and the hive. Workers from other hives lack these abilities and ignore cells with dead pupae. Rothenbuhler crossed the two strains and then performed some additional crosses with the hybrid generation. These experiments revealed that the threshold of each behaviour pattern was under the control of separate genes called U (for Uncap) and R (for Remove). Workers endowed with two copies of the recessive allele of each gene (uurr) would both uncap cells and remove the diseased pupae. Workers with one or two copies of the dominant allele of each gene (UURR or UURr or UuRR or UuRr) would do much less of either activity. Through his mating experiments Rothenbuhler produced some bees with the genotype Uurr. These failed to uncap cells, but they would remove diseased pupae provided Rothenbuhler lifted off the wax cell caps.

2.7 The unit of natural selection

So far it has been implicitly assumed that behaviour functions for the benefit of the individual performing it; but it would be more accurate to state that behaviour functions for the benefit of the descendants of the individual performing it. This is because individuals die but descendants, hopefully, exist for ever.

Darwin was brilliant enough not only to propose the theory of natural selection, but also to consider what at first appears to be a crushing counter-example: the existence of sterility in many social insect worker castes. Surely, it might be argued, such castes cannot have evolved by natural selection because the bearers of traits associated with sterility leave, by definition, no offspring. However, Darwin realised that sterility in such circumstances could evolve by a process he termed 'family selection'. For example, he pointed out that 'breeders of cattle wish the flesh and fat to be well marbled together; the animal has been slaughtered, but the breeder goes with confidence to the same family'.

Family selection is nowadays termed kin selection and is considered in more detail in Chapter 4. Here, still in the context of the unit of selection, it may be worth mentioning that instances occur when an individual's behaviour is to the individual's detriment, but to the advantage of an individual of another species! Flukes of the genus *Leucochloridium* are parasites of certain snails. At some point in their life cycle, the flukes invade the horns of their host snails where they can be seen through the skin, conspicuously pulsating. This tends to make birds – who are the next host in the life cycle of the fluke – bite off the tentacles, mistaking them for insects, it has been suggested. What is interesting here is that the flukes seem also to manipulate the behaviour of the snails, changing it with respect to light. The normal negative phototaxis is replaced in infected snails by positive light-seeking. This carries them up to open sites where they are presumably more likely to be eaten by birds, and this benefits the fluke, but obviously not the snail. Such an example should caution us in our eagerness to postulate functions for behaviour.

3 Learning

The previous chapter discussed the evolution of behaviour. It concentrated on behaviour that is usually thought to have a genetic basis. Obviously, however, much behaviour is all or in part dependent on learning. Learning is something most people think they could easily recognise. Nevertheless, producing a definition of learning is not easy. Kimble, in a well-known textbook, wrote that 'learning refers to a more or less permanent change in behaviour which occurs as a result of practice'. Thorpe defines learning as '. . . that process which manifests itself by adaptive changes'. It is important to remember that, as discussed in Chapter 1, a combination of both instinct and learning may equip an animal with a set of adaptive responses to the environment. It is also worth remembering that learning is a process and therefore cannot be measured directly. What is measured is the outcome of that process; in other words, what is learned.

Thorpe put learning into six categories, which will be used here. The six categories are: habituation; classical conditioning; operant conditioning; insight learning; latent learning; and imprinting. Animals vary in how much they are capable of learning. In general, the further up the phylogenetic tree they are, the more they are able to learn; so, for example, birds can learn more than insects and cats more than birds. As will be discussed in later chapters, man has an enormous capacity for learning which gives him rather special characteristics.

3.1 Habituation

If someone taps on a table top as a snail is crawling along it, the snail will immediately withdraw into its shell. Within a short time the snail will re-emerge and continue on its way. A second tap will also be followed by withdrawal, but the snail will then re-emerge more quickly than the first time. After several taps the snail will no longer withdraw at all but continue to crawl uninterrupted. The snail is then said to have **habituated** to the stimulus of a tap on the table top.

However, there could be other explanations for ceasing to respond apart from habituation. Repeated exposure to a particular stimulus can reduce an animal's sensitivity to that stimulus. This is known as **sensory adaptation**. In the case of the snail, the relevant nerve endings may simply have ceased to pick up the vibrations caused by the tap. On the other hand, the animal may continue to sense the stimulus but fail to respond because of either **muscle fatigue** or a change in **motivation**. The animal which is no longer hungry may ignore food but this does not necessarily mean it is unaware of it.

A series of experiments has been done with the sea snail, *Aplysia*, which shows how these other explanations may be ruled out. Touching the mantle shelf of a sea snail is followed by gill withdrawal. After repeated touching the gill ceases to withdraw. Habituation has occurred. However, if the mantle shelf is touched at the same time as some other part of the body is tapped, the gill withdraws again. This suggests that the original cessation of withdrawal cannot have been due to muscle fatigue or sensory adaptation. Neurophysiological studies have confirmed this. For example, recordings from the afferent nerves while touching the mantle shelf show activity even when gill withdrawal has ceased. Direct electrical stimulation of the appropriate efferent nerves will result in gill withdrawal even after habituation. This implies that habituation occurs as a result of changes in the animal's central nervous system, rather than at a peripheral level, either sensory or muscular.

Habituation occurs in all animals, and is not normally permanent. The appropriate response will reappear over time. However, in mammals, although a response reappears in the short term, it may differ from the original. There may also be long-term changes in the animal's behaviour. For example, most animals are startled by sudden noises. If the same noise is repeated they rapidly habituate to it and cease to show a startle response, such as looking around. The reappearance of the noise an hour or so later will again produce the startle response but it may be less obvious than before and the animal will certainly habituate more rapidly.

It is easy to see the possible function of habituation. An animal needs to respond appropriately to changes in its environment. The rabbit that scuttles down its hole whenever a tree rustles will have no time to feed. Equally, the rabbit that does not disappear quickly when it hears a strange noise runs the risk of being caught by a predator. The likely function of habituation is to distinguish between familiar and novel stimuli. Animals are likely to become habituated to common background noises, such as the wind, but will still react to infrequent, strange sounds. Habituation is important in the development of behaviour. For example, by this means the young animal learns to recognise the signs of danger, and not to react to every movement and shadow. Man also shows habituation. After a while we do not 'hear' the constant ticking of a clock or notice the nasty smell in the kitchen. Young children will continue to sleep while there are familiar, often loud, noises all around them but will wake up at a quieter but strange sound.

3.2 Classical conditioning

During habituation an animal simply learns not to respond to a particular stimulus. In most other forms of learning an animal learns to associate one stimulus with another. These forms of learning are therefore often called **associative learning.**

At the turn of the century a Russian, Ivan Pavlov, conducted a series of experiments with dogs. As a result of these he described a type of associative learning which is known as **classical conditioning**. When a hungry dog is presented with food it salivates. Pavlov placed his dogs in a piece of apparatus which rendered them almost immobile and also enabled him to collect saliva through a tube inserted in their salivary glands (Figure 3.1). Just before the dog

Figure 3.1 Apparatus used by Pavlov in his conditioning experiments. During the experiments a light (the conditioned stimulus) appears in the window and meat powder (the unconditioned stimulus) is delivered to the food bowl.

was given food Pavlov would ring a bell. After a while the dog salivated when it heard the bell even if it was not then given food. In other words, the animal had learned to associate the sound of the bell with the arrival of food and eventually to respond to the bell in anticipation of the food. In Pavlov's terminology, the food is called the **unconditioned stimulus (UCS)**. This produces the **unconditioned response (UCR)**, or natural response to the UCS, salivation. Over time the bell, called the **conditioned stimulus (CS)**, will elicit salivation without food. Salivation has then become the **conditioned response (CR)**.

Classical conditioning can, therefore, be seen as learning to associate one event with another, simply by observing the two to be associated. In the early days of the investigation of this form of learning it was assumed that as long as the conditioned stimulus rapidly followed the unconditioned stimulus learning would occur. However, it is now clear that the process is not as simple as this. Simple association of the two stimuli does not guarantee learning.

In experiments with rabbits it has been shown that even if a light, acting as the CS, is only followed half the time by a puff of air, the UCS, the light will still come to elicit an eye blink. If one of two tones is played at the same time as the light, then all three stimuli will come to elicit the eye blink. However, if – maintaining the scheme of only giving a puff of air half the time – one tone is always followed by the puff of air but the other is not, the rabbits will learn to respond only to that tone and not to the light (Figure 3.2). This is not very surprising and makes good sense if the rabbit is seen as learning the stimulus which is most likely to predict the occurrence of the puff of air. In this case, this is the tone. Even in the situation where the light only predicted the occurrence of the puff of air half the time, as described earlier, it is obviously better for the rabbit to respond half the time than not at all.

Thus, classical conditioning is one way in which the animal learns to predict its environment. It learns what events are associated with other events. If it has

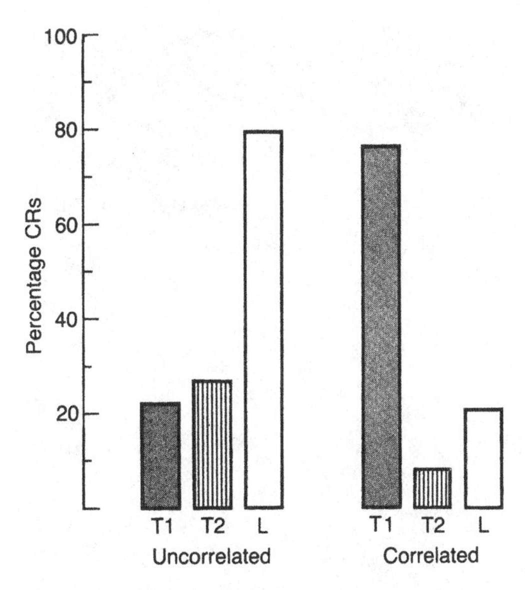

Figure 3.2 Graph to show the rate of responding of rabbits to two tones (T1 and T2) and a light (L), following conditioning to the combination of T1–L and/or T2–L. In the uncorrelated condition, both T1–L and T2–L were reinforced on a random 50% of the trials; in the correlated condition, T1–L was always reinforced and T2–L never.

learned to recognise a signal for a particular stimulus it may act in anticipation of the stimulus or of similar stimuli. This may be important for its survival. For example, rats confronted with a new kind of food will only take a little. If they become sick within a few hours they will not eat that particular food again. The taste of that food has become associated with being sick.

It is easy to find examples of human behaviour that can be explained in terms of conditioning. It has been suggested that some phobias can be explained in this way. For example, if someone does actually have a nasty experience while in the presence of a spider, then fear, and all the physiological responses that go with it, such as palpitations and sweating, may become associated with spiders. Fear becomes a conditioned response. Some people are prepared to account for much of human behaviour in terms of simple conditioning. They argue, for example, that money and poker chips derive their power to motivate people because of the associations with what they can buy. On a more mundane note, we may even start salivating when we see a sign for a restaurant, or feel a tremendous surge of emotion when we hear music associated with a loved one.

3.3 Operant conditioning

In the early part of the century Thorndike conducted a number of experiments using what he called 'puzzle boxes'. An animal, such as a cat, placed in the box could only get out to reach food by pressing a catch or panel which released the door. In the late 1930s Skinner developed what has now become known as a **Skinner box** in which the animal, usually a rat, has to press a lever to receive

Figure 3.3 Skinner box for use in operant conditioning the rat.

food (Figure 3.3). Skinner coined the phrase **operant conditioning** to describe this type of learning. It has also been called **instrumental learning** or **trial and error learning**, for reasons that will become clear.

When a rat is first put into one of these boxes it rapidly explores its surroundings. During the course of exploration it will accidentally press the lever. Food appears. Gradually the rat learns to press the lever to obtain food. However, the discovery of the association between pressing the lever and obtaining food is made by trial and error, hence the name trial and error learning. The rat has to learn that pressing the lever is instrumental in obtaining food, hence the name instrumental learning. In ordinary language, the rat learns to press the lever because it is rewarded by food for doing so. Experimental psychologists talk about the lever press being positively **reinforced** by food. An animal may also learn *not* to do something if the reinforcer is unpleasant, such as a mild electric shock.

One way of looking at the difference between classical and operant conditioning is to describe classical conditioning as the learning of an association between two stimuli, but operant conditioning as the strengthening of the association between stimulus and response (e.g. pressing the lever) by following the response with a reinforcing stimulus (e.g. food). There are many possible examples of operant conditioning in human behaviour. Children are constantly being rewarded for good behaviour and punished for bad, and as adults we may still alter our behaviour depending on whether other people show approval of it or not.

In the experimental setting an animal can learn as a result of a large number of different **schedules of reinforcement**. These are the patterns of food presentation chosen by the experimenter. Speed of learning depends on the schedule of reinforcement, as does the time the animal takes to stop pressing the lever if the reinforcer, food, is no longer delivered. For example, with a schedule of **partial reinforcement** (i.e. the animal only receives food for some, not all, lever presses) the animal will take longer to learn to press the lever than if it receives food all the time, but will stop pressing the lever less quickly if the food is withdrawn completely. An example of human behaviour that can be explained in this way is the obsession that people can develop about playing fruit machines. Fruit machines only pay out money some of the time; in other words, they provide partial reinforcement. This means that, even if they have not paid out for a very long time, the players find it very difficult to stop.

3.4 Insight learning

Classical and operant conditioning can be seen as simple associative learning. A number of psychologists have argued that there are other more complex types of learning. One of these is known as **insight learning**. This occurs when an animal solves a problem by apparently looking and assessing the situation. People say that they feel 'the penny drops' or the solution 'comes in a flash'.

The first experiments to demonstrate this type of learning were done by Köhler with chimpanzees. A typical problem they had to solve was to obtain food that was just out of reach outside their cage. The chimps were provided with sticks of various lengths. In order to reach the food they had to fit two sticks together and then pull the food towards them. A number of the chimpanzees were able to do this. Köhler saw this kind of learning as a product of perceptual reorganisation. An animal learned to solve the problem when it learned to see the various elements of the problem in a new way.

However, it can be argued that this kind of learning may be explained in terms of operant conditioning. Young chimps often play with sticks. During the course of this play they will fit the sticks together and poke at objects. They may, therefore, have already learned during play those responses necessary to obtain the food outside their cages. Applying this behaviour to solving the problem of obtaining the food then becomes similar to operant conditioning.

3.5 Learning sets

The explanation of insight learning in Köhler's chimps in terms of operant conditioning is not entirely satisfactory. However, the chimps did at least demonstrate that previous experience may influence what is learned. This is also demonstrated by the work of Harlow on problem solving in rhesus monkeys. He showed that monkeys may learn a strategy or rule, **learning set**, which is then applied to all similar problems, rather than each problem being considered individually.

Harlow presented his monkeys with two objects which were placed over two food wells, for example a red block and blue block. The monkey then had to learn which object hid the food. Harlow gave his monkeys hundreds of these problems to solve with different objects to choose from. Initially the monkeys took several trials to learn which object concealed the food but after a large

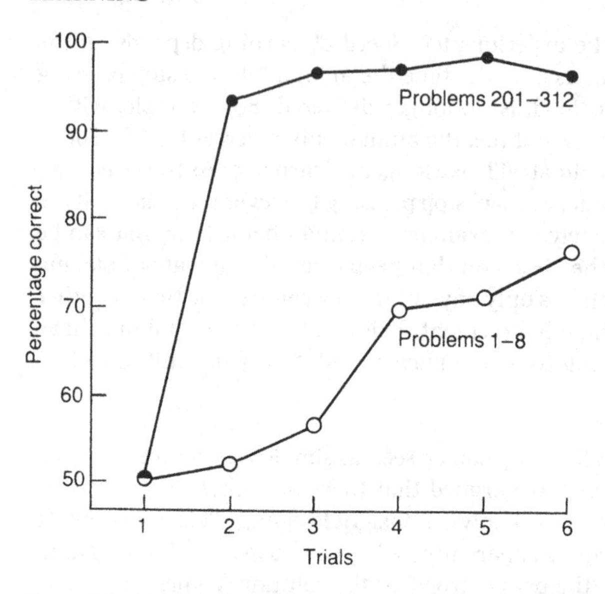

Figure 3.4 Graph to show the performance of rhesus monkeys over the first six trials of a series of two-choice discrimination problems. The lower curve represents performance on the first 8 problems, the upper curve represents performance after training on 200 problems.

number of problems they nearly always chose correctly after one trial (Figure 3.4). The monkeys' solution to the first problem can be described in terms of conditioning. The appropriate object becomes associated with food and therefore elicits approach. After solving many problems the monkeys may develop a strategy, for example: 'If food is found, stay with that object, if not choose the other'. Applying this rule they need only one trial to solve the problem. In forming a learning set, the monkey characterises the solution of a problem in a different way from the characterisation he makes during conditioning – in terms of a general rule, rather than a set of individual associations. Children who learn how to add or subtract, for example, are acquiring a learning set. They do not have to learn every individual sum that they might come across. It is this complexity of representation that may distinguish simple conditioning from higher order learning.

3.6 Latent learning

A rat which has explored or been taken through a maze, before being required to solve it to obtain food, will usually find the food faster than a rat which has had no experience of that maze. The implication of this finding is that the rat has learned something about the maze while exploring but without any immediate reward for doing so. What is learned remains hidden or 'latent' but may be of use at a later date, hence the term **latent learning**. Obviously, animals in the wild do learn about the geography of their home area. This knowledge may mean the difference between life and death when a predator strikes. The possible existence of latent learning has caused controversy amongst psychologists. Those who argue that learning can only occur as a result of reinforcement cannot concede that latent learning occurs at all. However, it has been shown recently that seeing something new is rewarding in itself. The rats going through the maze would therefore be reinforced just by the new experience.

3.7 Imprinting

A particular form of learning often given a unique status is **imprinting**, first made famous by Konrad Lorenz. A number of birds and mammals are born sufficiently physically mature to be able to move about freely almost immediately after birth. These species are said to be **precocial**. Precocial young normally become attached to their mothers and follow them closely after the first few days of life. This process is known as imprinting, or more precisely **filial imprinting**. Although in the wild the young usually become imprinted with their mother, they can become attached to a wide range of stimuli. Konrad Lorenz first observed the phenomenon when a group of greylag goslings became imprinted with him and subsequently followed him around. Later experimenters have demonstrated the effect with, for example, flashing lights, patterned boxes, buckets and members of other species.

It is now known that there is a period called the **sensitive period** when imprinting occurs most strongly and easily in the young, precocial bird. However, the length of this period is not fixed and varies between individuals of the same species. The beginning of the period is dependent on the development of the bird's visual and auditory systems. Obviously the bird cannot follow an object until it can distinguish that object from others. The end of the period is dependent on the bird's experience during this time. Chicks reared in groups cease to follow novel moving objects within three days after hatching, whereas those reared in isolation remain responsive much longer. If a bird is forced to live with a different object to that to which it has become attached, after the apparent end of the sensitive period, then it may eventually prefer this to its original favoured object.

Much of the research on filial imprinting has concentrated on birds. However, this process can also be observed in mammals such as goats, sheep and some antelopes. All show a tendency to become attached to and follow their mothers soon after birth, and then to prefer her to other animals. One aspect of the adaptive significance of filial imprinting is clear. It helps to ensure that the young stay close to their mother. However, more investigation, particularly in the wild, is needed before the process is fully understood. For example, in addition to their mother, young precocial birds also become attached to other members of the brood, and mothers become attached to their young. It is not clear how important these factors are for keeping families together, relative to filial imprinting.

As birds develop, the strength of filial imprinting declines. However, its effects may last longer. Lorenz and others have observed that birds imprinted with females of another species, or indeed humans or cardboard boxes, later attempted to court and mate with them. These birds had become **sexually imprinted** with these objects. This is demonstrated by the systematic cross-fostering experiments of Immelmann. He placed a single egg of one species, e.g. zebra finch, in the clutch of a second, e.g. Bengalese finch. When mature, male zebra finches brought up in this way always showed a preference for mating with Bengalese rather than zebra finch females. Female zebra finches also showed a preference for Bengalese male finches, though not to the same extent. Thus, imprinting, which has the more obvious adaptive significance for the young of the species, also has later consequences.

3.8 Constraints on learning

Animals cannot learn everything equally well. There are constraints on what they can and will learn. For example, rats can learn to avoid a visual cue associated with shock and to avoid certain tastes if they are made to vomit later. However, it is almost impossible to teach a rat to avoid certain tastes if they are paired with a shock or to keep clear of particular visual cues if they are paired with induced vomiting. Constraints, therefore, exist on which rewards or punishments can be used to change behaviour. These constraints usually make adaptive sense. It is obviously useful for rats to be primed to remember tastes that make them sick. Food is more likely to do that than the sight of a particular object. Thus, animals may be predisposed to learn those things which are most useful to them. It is very difficult to teach an animal to associate two events unrelated to events which could have adaptive significance, or to teach it to perform some act for a reward which is not in some way similar to its natural repertoire of behaviour. For example, it is very difficult to teach a rat to groom itself in order to avoid an electric shock since this would not make sense in its natural environment.

3.9 Memory

All learning implies the existence of **memory**. If nothing were left from previous experiences nothing would have been learned. Most studies of memory involve humans, and many of these studies make use of lists of words or nonsense syllables. It is much easier to ask people what they remember than to try to assess memory purely behaviourally. Most of the following discussion will, therefore, be about memory in humans. However, there is no reason to suppose that memory, at least in the higher animals, does not work in a similar way.

Memory can be viewed in terms of three stages: **encoding**; **storage**; **retrieval**. During encoding what is seen, heard, smelled, etc. is transformed into a form suitable to be processed by the memory system. The encoded information is then stored in some way. This information must later be retrieved when needed. A faulty memory could involve a failure at any one of these stages.

There are at least three different kinds of remembering. **Recall** is the kind of memory most easily tested in the laboratory. It is simply an active recall of something learned or experienced in the past, for example, a list of words or how to solve a problem. In **recognition** we simply realise that something is familiar and that we have come across it before. The final way to show that some memory of the past exists is to demonstrate that some previously learned material can be learned more quickly than it could be if it were completely unknown. For example, if we have some knowledge of a foreign language, it is easier to pick it up again at a later date than if we have never learned it at all.

Three explanations are commonly given for why we forget things. These are: decay of memory through disuse; interference effects; motivated forgetting. Decay through disuse assumes that learning leaves a 'trace' in the brain. As time passes the normal metabolic processes in the brain are assumed to cause this trace to fade, resulting in the memory disappearing. Although this sounds plausible, no direct evidence exists to support it. One problem is that we do remember some things for a very long time (such as how to ride a bicycle) but other things (such as a poem learned for class) we forget very quickly. It is difficult to know why the trace should fade for some memories and not for others. Another argument against the decay theory is that we do recover memories we have apparently lost. For example, old people often remember things from their childhood they had thought they had forgotten. One problem for the decay theory is that it is not known precisely what, if any, biochemical changes take place in the brain when something is learned. Until this is known it is difficult to assess the likelihood of the trace decaying. It is quite plausible that organic changes in the brain over time may have some effect on memory though this is obviously not the only explanation for forgetting.

Another explanation for forgetting suggests that it is not so much the passing of time that affects memory but what we do or try to learn before or after learning something that affects its retention. If new learning appears to interfere with what has already been learned, this is known as **retroactive interference**, because the interference is working backwards. This can be demonstrated by comparing retention of learned material after sleeping, with retention after a period of waking. As can be seen in Figure 3.5 more is lost from the memory after being awake than after a period of sleep. What is experienced and to some extent remembered while awake has interfered with the memory of the earlier material. The opposite type of interference of memory may also occur. This is known as **proactive interference**. Here, previously learned information hinders the retention of new information. For example, if a friend moves house we may have difficulty in remembering the new address because the old one keeps springing to mind.

Sometimes, forgetting is not the result of physiological processes or interference but occurs because a person actually wants to forget. This is known as **motivated forgetting**. For example, people often forget all the circumstances surrounding a very traumatic event. However, although these people are unable to recall the event consciously, the circumstances may reappear in a dream or be recalled under hypnosis, thus demonstrating that the memory is not completely lost.

There appear to be two types of memory in humans; **long-term memory** and **short-term memory**. Short-term memory may simply be used as a short-term store or as active memory, for example when we remember a telephone number just long enough to dial it. Information is first stored in the short-term memory. If it is not rehearsed or repeated and transferred to long-term memory

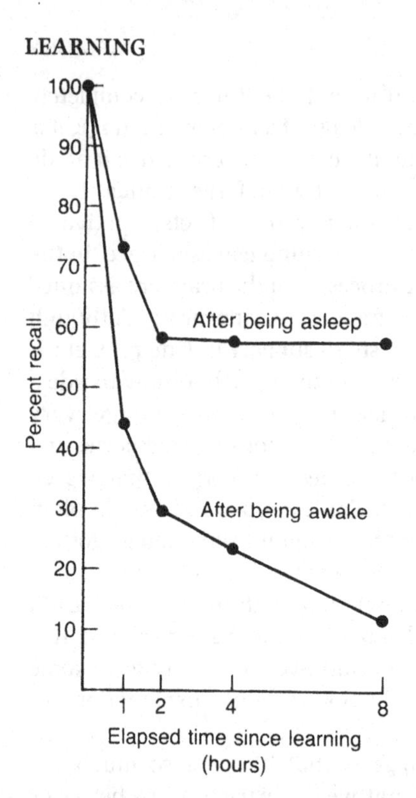

Figure 3.5 Graph to show the retention of information when the subject is either awake or asleep during the interval between initial learning and later recall test.

it will rapidly be forgotten. Once lodged in long-term memory it may be retained for ever. Loss of information from the short-term memory could be as a result of trace decay whereas loss from long-term memory is more likely to be a result of retrieval problems or improper storage, such as interference effects.

3.10 Intelligence

One dictionary definition of intelligence is 'the capacity to acquire and apply knowledge'. Knowledge itself is not necessarily the mark of intelligence. A high degree of intelligence also implies an ability to learn and solve problems. Animals higher up the phylogenetic scale tend to have a greater ability to learn and thus have traditionally been regarded as more intelligent than those lower down. Language provides man with an enormously useful tool for learning and therefore, quite apart from increases in any other aspects of learning ability, considerably boosts his intelligence in comparison with other animals. Language provides an excellent code for storing information. It also enables information to be passed easily from one person to another. A full consideration of the development and measurement of intelligence in man, however, is beyond the scope of this book.

There are at least three obvious evolutionary pressures towards greater intelligence in animals. First, is the advantage conveyed on an animal capable of adapting to a changing environment; this is, of course, achieved by learning new behaviour. Second, there is the pressure to develop ways of outwitting predators; again, this involves the development of the capacity to learn. Finally, there is the pressure to adapt to society itself, to be increasingly competent in dealing with other members of that society; this involves a heightened ability to predict and manipulate the behaviour of others. In comparison with insects, individuals' social roles in primate societies demand a high degree of learning. This is taken to the extreme in human society where man's intelligence has resulted in and is affected by the enormously complex social relationships with which he is required to cope. The structure of human society and the development of social relationships will be discussed further in Chapters 7 and 8.

4 Co-operation and competition

4.1 Territorial behaviour

As much has been written about the definition of the word 'territory' as about the definition of perhaps any other biological term. Most ethologists would be happy with the suggestion that a territory is 'a more or less exclusive area defended by an individual or group'. This definition includes three important points. First, a territory allows its owner certain **rights** (more or less exclusive use). Secondly, territories are **defended** – albeit sometimes subtly. Thirdly, territories are **areas** – they are relatively fixed in space.

Modern research into territorial behaviour tends to centre on the concept of **economic defensibility**. This states that we would only expect an animal to spend time and energy interacting with others to defend a territory when this yields greater benefits than an alternative behaviour, for example ignoring others in the population and spending the whole time exploiting a shared resource (see Figure 4.1).

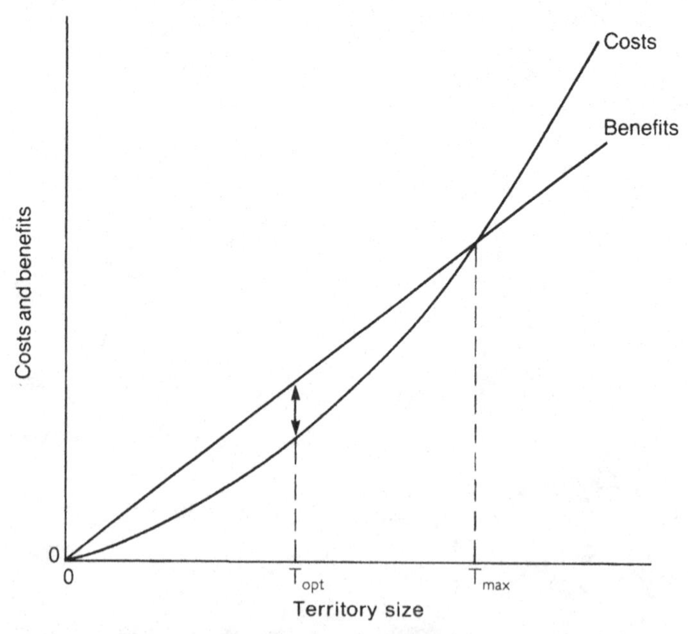

Figure 4.1 Hypothetical relationships between territory size and the costs and benefits of having such a territory. The maximal predicted territory size (T_{max}) occurs when the costs exceed the benefits. The optimal predicted territory size (T_{opt}) is given when the difference between the benefits and costs are a maximum. For such an analysis to be valid the costs and benefits must be measured in the same currency, for example energetic costs and energetic benefits.

Ecological factors favouring territoriality

Four main factors will influence the economic defensibility of a resource. These are: resource quality; resource distribution in space; resource distribution in time; and competition for the resource.

The range an animal will have to occupy to satisfy its requirements will depend on the abundance and distribution of the required resource. At one extreme, if the resource is of poor quality and sparsely distributed (e.g. dry grass eaten by wildebeest on an African savannah) the animal will have to roam over a large area and it is unlikely that it will be able to defend this economically. On the other hand, high-quality food may be worth defending even if fairly evenly distributed (e.g. succulent buds browsed by small forest antelopes).

If food patches are ephemeral, then individuals may have to roam over a large area to exploit sufficient patches to stay alive. In such a case the best option for an individual might be to live non-territorially at the centre of distribution of the patches in order to minimise its travel time when foraging. If resources are more predictable in time, for example because they are renewed at a sufficient rate for an individual to exploit the same patch for long periods, then it may be advantageous for animals to set up territories.

The number of competitors and individual differences in competitive ability obviously influence the economics of territory defence. For example, large piles of high-quality food are usually not defended (e.g. finches feeding on substantial clumps of seeds) as the competitor pressure would presumably be too great. Smaller piles of food may be defended territorially, however.

Benefits of territoriality

There are many benefits of territoriality (see Table 4.1). Here, one example is considered in detail: the winter feeding territories of the pied wagtail, *Motacilla alba*, on a meadow in the Thames Valley, as studied by Davies and Houston. In this locality, some pied wagtails defended territories along a river while others fed in flocks on flooded pools nearby. The territory-owning birds exploited a renewing food supply, namely small insects which were washed up by the river on to the muddy banks. Owners typically walked a regular circuit around their territories, up along one bank and then back down the other side again (Figure 4.2[a]). By systematic depletion of the food supply on the territory the owner can, in effect, make it unprofitable for intruders to land. The most profitable place to feed is just ahead of the owner, as there the prey has had the most time for renewal. If an intruder landed there it would be easily spotted. If, on the other hand, the intruder landed behind the owner it would be feeding over stretches the owner had recently depleted. Measurements showed that intruders often fed at an unprofitable rate, even if they managed to sneak on to the territory undetected, precisely because they fed over depleted stretches. This explains why intruders were usually very noisy when flying over a territory, calling loudly 'chisick'. When this happens the owner, if in residence, replies 'chee-wee', whereupon the intruder usually departs. Perhaps the intruder's noisy calls are an enquiry as to whether an owner is present. If one is, then this signals a depleted food supply so it pays the intruder to move on.

Table 4.1 Examples of the benefits of possessing a territory

Species	*Benefit*	*Source*
MOLLUSCA		
Owl limpets (*Lottia gigantea*)	Food	Stimpson, 1970
ARTHROPODA		
Honeybees (*Apis mellifera*)	Decreased predation on hive or nest	Butler, 1974
TELEOSTII		
Three-spined stickleback males (*Gasterosteus aculeatus*)	Attract females	Tinbergen, 1951
	Decreased predation on eggs in nest by conspecifics	Black, 1971
AMPHIBIA		
Bullfrog males (*Rana catesbeiana*)	Space for sexual display	Emlen, 1968
REPTILIA		
Galapagos marine iguanas (*Amblyrhynchus cristatus*)	Males: space for sexual display Females: egg nest-site	Eibl-Eibesfeldt, 1966
AVES		
Rufous hummingbirds (*Selasphorus rufus*)	Food	Kodric-Brown and Brown, 1978
Great tits (*Parus major*)	Food, decreased predation	Krebs, 1971
	Males: attract females	Dunn, 1977
MAMMALIA		
Belding's ground squirrel (*Spermophilus beldingi*)	Decreased cannibalism of young by conspecifics	Sherman, 1981
Rhesus monkeys (*Macaca mulatta*)	Food	Neville, 1968

The wagtails maintain the same length of territory, a circuit of on average 600 m, throughout the whole winter, despite evident changes in resource abundance. Maybe continuous adjustments in territory size to track daily optima would be too costly. At the start of winter, neighbours spend a lot of time in boundary disputes, but once settled these boundaries are respected and are maintained simply by short displays at intervals throughout the day. Instead of daily changes in territory size, owners adjust to variations in food abundance by changes in their behaviour towards intruders and changes in the amount of time they spend on their territories.

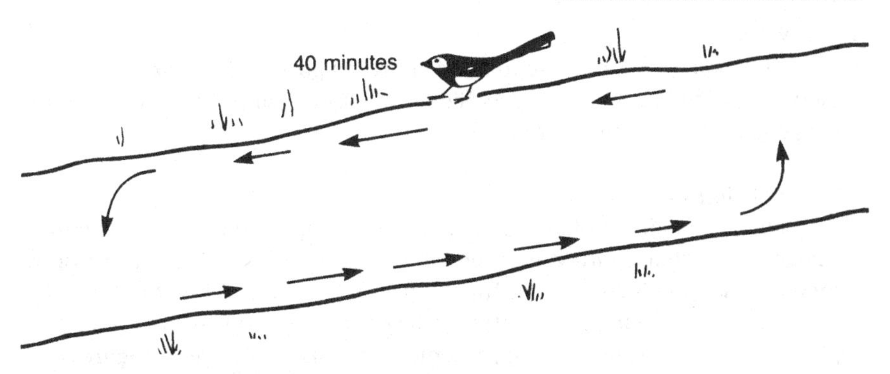

Figure 4.2(a) Pied wagtails exploit their feeding territories systematically. A complete circuit, at Davies and Houston's study site, took about 40 minutes.

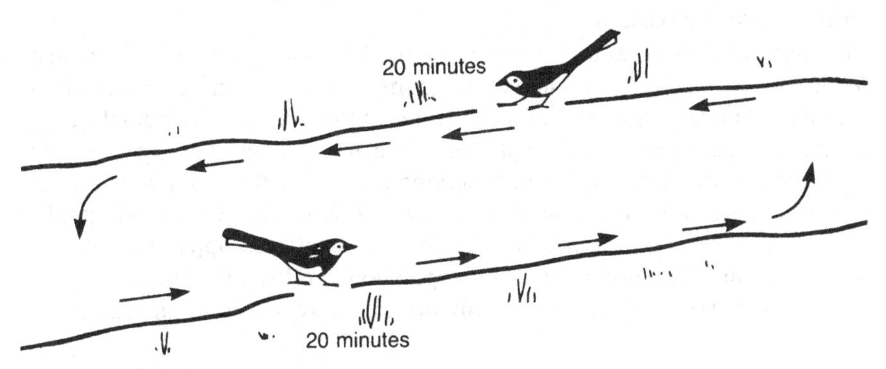

Figure 4.2(b) When a territory is shared between two pied wagtails, each walks approximately half a circuit behind the other, and so crops only about 20 minutes' worth of new food.

Four levels of territory defence and time allocation can be observed.

1 When food is scarce on the territories, the owners leave and feed elsewhere but keep returning at intervals to announce ownership and evict intruders. This suggests that territorial defence is a long-term investment and that a territory is worth retaining even through periods of low prey abundance.
2 At intermediate levels of food abundance, owners spend all day on their territories and evict all intruders.
3 If food increases further, however, an owner begins to share its territory with a subordinate individual, a **satellite**, usually a first-winter juvenile or an adult female. Sharing the territory brings costs to the owner because the satellite depletes the food supply on the territory. Owners and satellites usually share the territory by each walking half a circuit behind the other (Figure 4.2[b]). Sharing also brings a benefit, however, because the satellite helps with defence.
4 At very high levels of food abundance, owners abandon all attempts at defence. This makes good economic sense because with such high levels of food abundance, for example when there is a sudden emergence of insects in the spring, an owner's feeding rate is hardly affected by the presence of other birds.

4.2 Aggression

Aggression, sometimes referred to by the phrase **agonistic behaviour**, serves a variety of functions. No single classification is universally approved, but the arrangement given below is typical.

Territorial aggression

The territory holder utilises dramatic signalling behaviour to repulse intruders. Escalated fighting may be employed as a last resort. The losing contender has submission signals that help it to leave the field without further physical damage. Individuals entering territories held by members of the opposite sex may use elaborate appeasement signals to change the aggressive displays of the territory holder into courtship.

Dominance aggression

The aggressive displays and attacks mounted by dominant animals against fellow group members are similar in many respects to those of territorial defenders. However, the object is less to remove the subordinates from the area than to exclude them from desired objects (e.g. food, mates, sleeping sites) and to prevent them from performing actions for which the dominant animal claims priority. In some species dominance is characterised by special signals, such as the leisurely stroll with head held high and tail erect apparent in rhesus macaques, and the particular facial expressions and tail posture of wolves. Subordinates respond with an equally distinctive repertoire of appeasement signals.

Sexual aggression

In some species males may threaten or attack females for the sole purpose of mating with them or forcing them into a more prolonged sexual alliance. An extreme example is the behaviour of male hamadryas baboons, who recruit young females to assemble a harem and continue to threaten and harass these consorts throughout their lives in order to prevent them from straying.

Parental disciplinary aggression

The parents of many species of mammals, where the young may rely on parental protection, comfort and/or feeding for months or years, direct mild forms of aggression at their offspring to keep them close at hand, to urge them into motion, to break up fighting, and so forth.

Weaning aggression

There exists a period in the life of a growing animal during which its fitness is best served by the receipt of more parental care, while the fitness of its investing parent (often the mother) is best served by the termination of parental care, so that the parent can get on and invest in other offspring. During this period the parents may threaten or even attack, though usually gently, the persistent offspring.

Moralistic aggression

The societies of higher organisms – particularly the social insects, social carnivores and primates – operate by the general acceptance of 'rules' or 'codes of behaviour'. Transgressions of such rules may be punished.

Predatory aggression

There is some disagreement as to whether predation should be considered as an example of aggression. At any rate, it differs (along with antipredatory behaviour, mentioned below) in that it is usually **interspecific**, whereas the above six categories (with the occasional exception of territorial aggression, as when two species hold interspecific feeding territories) are **intraspecific**.

Antipredatory aggression

Usually this serves to startle the predator, or warn other prey individuals. Occasionally, however, predators may be injured or even killed.

Aggression in humans

In accounts of human aggression we find war, riot, various individual acts from homicide to hooliganism, and many other phenomena. Yet each item on this list itself includes a variety of aggressive acts. War may include hand-to-hand combat or the impersonal bombing of a target. Indeed, in many of its aspects modern warfare involves no aggression in the sense of interpersonal violence: instead, the modern warrior operates weapons by remote control.

At the other extreme are the types of aggression seen professionally by many psychiatrists. Socially unacceptable violence may reflect a toxic psychosis induced by drugs, or other mental disturbances. Possibly none has any counterpart in animal conduct. In hoping to achieve an understanding of human violence, ethological insights may be helpful, but a biological frame of reference is unlikely to be adequate in itself. Studies of animals may permit one to make certain generalisations about factors that increase the probability of aggressive episodes, but even though aggressive behaviour in humans has presumably evolved under the influence of natural selection, it is clear that our biological nature and aggressive instincts are inadequate by themselves for harmonious coexistence in the complex societies that have only developed in the last 10 000 years.

Part of our problem is that the ever-increasing efficiency of artificial weapons means that less and less effort is required to carry out a damaging attack. Additionally, we have steadily improved the range at which our weapons can kill. This progression presumably began with throwing a missile instead of hitting directly. With the use of arrows in conflict, rather than just in the hunting of prey, the distance between combatants increased still further. When gunpowder was invented bullets could be sent to kill at distances at which it was impossible to make out the details of the victim. This added an impersonal element to fighting, greatly reducing the possibility of appeasement signalling between combatants.

We will return to aggression in Chapter 6 in the context of how animals, including humans, communicate to reduce the risk of territorial invasion or injury to themselves.

4.3 Costs and benefits of leading a social existence

Just as an understanding of territoriality requires an appreciation of the costs as well as the benefits of territoriality, so an understanding of sociality necessitates a consideration both of the advantages and the disadvantages of group living. The name given to the discipline which deals with the systematic study of the biological bases of all social behaviour is **sociobiology**.

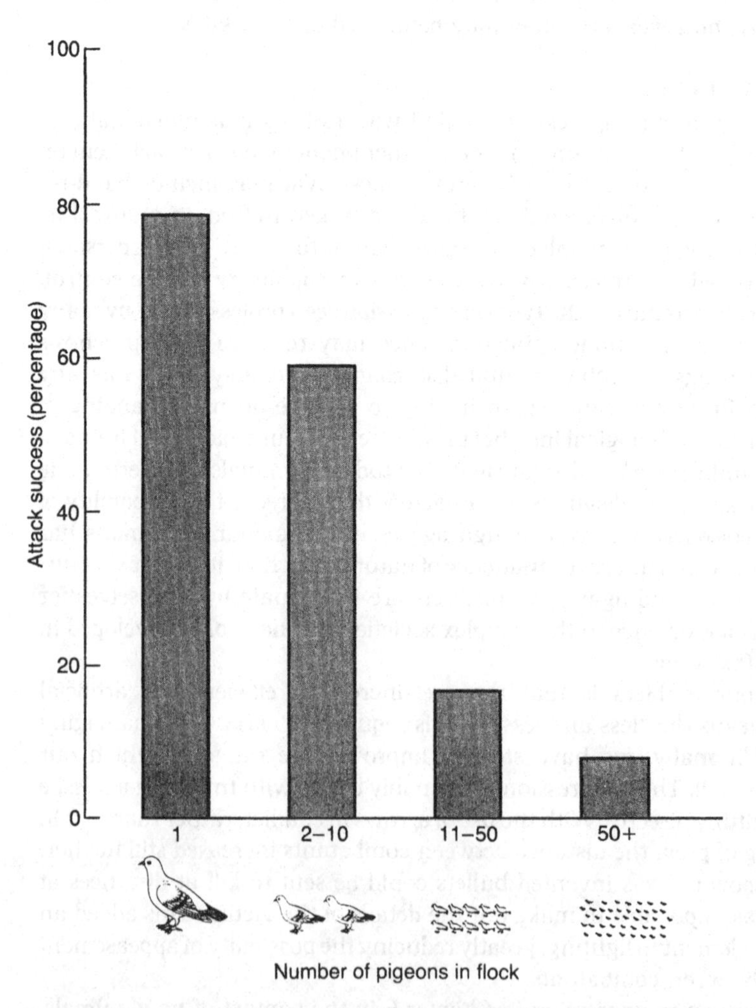

Figure 4.3(a) Larger flocks of woodpigeons are less likely to be attacked successfully.

Advantages to prey species by reducing predation

For prey species that depend for their safety on flight rather than camouflage or antipredatory aggression, early detection of the predator may mean the escape of the prey. It seems obvious that the more detectors there are, the higher the probability of such early detections. Kenward used a trained goshawk to attack feeding flocks of woodpigeons, and found that the predator's attacks were less successful when directed at larger groups of pigeons (Figure 4.3 [a] and [b]).

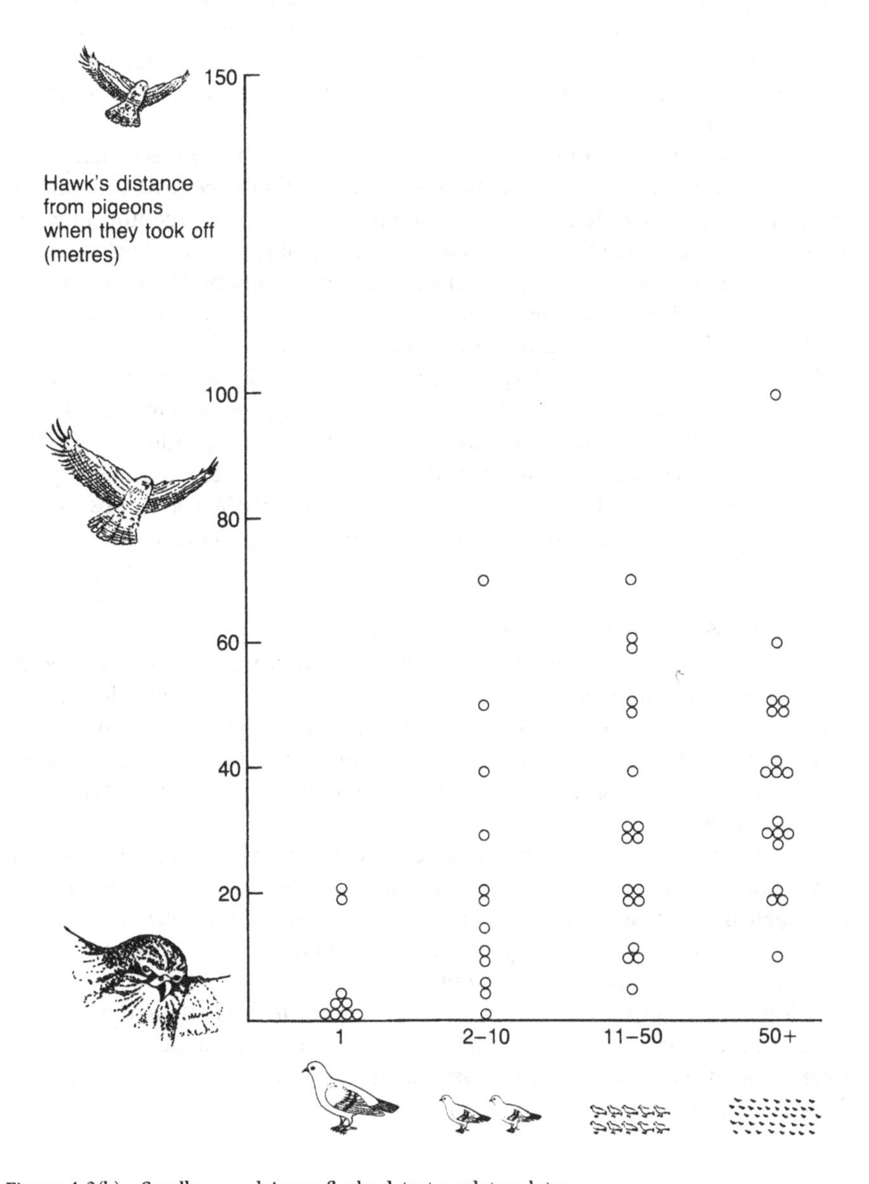

Figure 4.3(b) Smaller woodpigeon flocks detect predators later.

If the prey species is not a great deal smaller than the predator, or if it has particularly effective weapons, several of the prey species attacking in concert can sometimes deter or thwart the predator's attack. Wasps, ants and bees, of course, can defend their colonies very effectively against casual predators thousands of times their size. Musk oxen threatened by wolves gather into a defensive formation with an impressive array of horns facing the predator and with vulnerable animals in the middle. It is also possible that a predator may be deterred from attacking a group of prey by the risk of injuries incurred indirectly. Tinbergen noticed that when a peregrine falcon flies over an airborne flock of starlings, the starlings clump. He suggested that a peregrine diving at high speed might be severely injured if it struck any but its intended victim. Interestingly, starlings do not form such tight flocks when peregrines fly beneath them.

If a predator has been neither detected nor deterred, it still has to select one prey animal and attack it. It seems possible that if there are several potential prey animals which flee in unpredictable directions, a predator may be confused and less able to concentrate on just one of them. Good evidence for such confusion effects come from an experimental study by Neill and Cullen, who showed that such confusion caused lower success rates by squid, cuttlefish, pike and perch attempting to prey on shoals of small fish.

Even if predators attack large groups as successfully as small groups or individuals, it may still pay individuals to reside in large groups. Imagine a group of five prey individuals and another group of ten prey individuals. If the two groups are equally likely to be attacked, individuals in the larger group are twice as safe. This effect will only fail to lead to grouping if larger groups receive proportionately more attacks or are attacked proportionately more successfully.

Advantages to species in obtaining food

Ward and Zahavi suggested that the behaviour of a number of bird species when they gather into large flocks might enable them to exchange information about food sources. It would obviously be beneficial for a bird following a poor day's foraging to observe and accompany other individuals which appeared better fed. The successful bird might not be able to prevent such observation and following.

A number of social carnivores are able either to catch larger prey or to catch prey more often when hunting in groups. Classic studies in the African Serengeti have shown this to be the case for lions, spotted hyenas and wild dogs. In the same way, solitary wolves live largely on carrion or small prey whereas packs take adult moose and deer.

Being in a group also enables some predators to hold their own better against other predators. A single hyena is no match for a lion, but a pack of hyenas can drive a lion from a carcass. However, a pack of hyenas is no match for a pride of lions.

Other advantages in group living

Many of the advantages described so far apply whether or not the members of the group are of the same species. Mixed-species groups do occur. Olive baboons and impala are often found, for example, in large mixed herds. This may be because together they can detect predators better by combining their respective abilities. Impala have extremely sensitive noses and ears, while baboons possess colour vision and may adopt high vantage points.

Animals may benefit from grouping by an improved ability to resist the physical environment. Woodlice huddle and so are able better to survive desiccation. Communal roosting may reduce heat loss. Another aspect of the inanimate environment to be overcome is the resistance to locomotion. The shoal formation of some fish is such that each individual benefits from the movement of the others, and thus the drag on all of them is reduced. Aerodynamic advantages have been postulated for large birds such as geese which fly in a V-shaped formation.

Permanent group living may permit communal breeding (see section 4.4 below) and substantial amelioration of the environment. Termites, for example, build huge mounds with their own internal air-conditioning system that serves to maintain the colony at an optimal temperature. Another advantage of a stable group is that it may allow more complex social relationships to develop. Members may be able to recognise one another as individuals and learn from each other.

Disadvantages of group living

Within a group of animals potential mates are easier to find. Mates are also, however, easier for a dominant individual to monopolise. Males generally compete among themselves more than females do, because as Trivers has pointed out, by doing so males can increase their reproductive output more than females can (see Chapter 5). In many species competition among males results in a dominance hierarchy with males at the top doing most of the mating. This is obviously unfortunate from the point of view of other males. In some species competition among males is so strong that all the others are driven from the group by the dominant male as, for example, in langurs, in many other one-male troops of primates, in seals, and in a number of antelope and some rodent species.

Females, too, may compete for reproductive opportunities. In packs of dwarf mongooses, wild dogs and wolves, usually only one dominant female reproduces, managing either to prevent subordinate females from coming into oestrus – to inhibit them from mating – or to kill their offspring. The evolution of a system of inhibition of the reproduction of subordinates is probably helped by a combination of several factors. First, if the dominant female is able to produce herself as many offspring as the group could support; a wild dog bitch, for example, can produce 16 puppies in a litter. Secondly, if subordinates are unable to breed successfully on their own, so that they do better to remain in the group and perhaps outlive the dominant. Thirdly, if the dominant is able to exercise control over the subordinates. And fourthly, if the subordinates are closely related to the dominant, so that the selective pressure to compete is reduced through kin selection (see section 4.4 below). There is positive feedback at work here. Any monopolising of reproduction means that the offspring tend to be more closely related to one another. This in turn probably favours more co-operation through kin selection, and reduces still further the selective pressure on subordinates to compete for reproductive opportunities.

Being in a group means that there is potential competition for food among the members, even if that food was only captured because of the participation of the group. How substantial this competition is will depend on the ecology of the species concerned. The presence of even a few companions compels resident herbivores, such as rabbits, to travel further each day to feed. Food competition is particularly intense among the social carnivores. A companion at a carcass means a reduction in the amount of food available to the eater, to an extent which does not apply to a grazer.

4.4 Evolution of altruism

Darwin's ideas about 'family selection' were mentioned briefly in section 2.7. Darwin's suggestion was that sterility might under certain circumstances be favoured if an individual's sterility was compensated by the extra number of descendants surviving to his or her relatives. This indeed is what seems to happen in social insects. For example, within a honeybee colony, the sterile female workers are typically the full or half-sisters of the few fertile bees (males and females) that will give rise to future colonies. In a colony with tens of thousands of individuals, the loss of a few workers through the stinging of potential predators hardly matters. The safety and protection of future reproductive individuals is the paramount consideration.

Altruism occurs when an individual's behaviour decreases its individual fitness but benefits another individual. The name given to the type of altruism where the decrease in an individual's fitness is more than compensated for by the increased fitness of its relatives is now termed **kin selection**.

Kin selection

Some of the most dramatic examples of kin selection occur in the social insects. In the African termite *Globitermes sulfureus*, for example, members of the soldier caste are literally walking bombs (Wilson, 1978)! Huge paired glands extend from their heads back through most of their bodies. When they attack ants and other enemies they fire a yellow glandular excretion through their mouths; this congeals in the air and often fatally entangles the aggressive soldier termites themselves as well as their antagonists. The spray appears to be powered by muscular contractions in the abdominal wall, and the contractions are sometimes so violent that the gland explodes together with the abdomen, spraying the fluid and the soldier in all directions.

The first steps in the quantitative determination of just how much benefit relatives have to receive to compensate an individual's altruism were taken separately by the geneticists R. A. Fisher and J. B. S. Haldane. Fisher was concerned with the evolution of distastefulness, the process by which nauseous flavours have been evolved as a means of defence. The problem is that predators frequently only realise that an individual prey is distasteful once they have killed the prey and have begun eating it. If predators learn to avoid such distasteful prey, then the prey species clearly benefits, but the individual killed obviously does not. The problem is enhanced when **aposematism** occurs, that is, when distastefulness is associated with striking coloration, as in many butterflies. It has been demonstrated that such coloration sometimes attracts predators. Fisher realised that the gregarious habit of many aposematic prey supplied the possible answer:

> For although with the [solitary] adult insect the effect of increased distastefulness upon the action of the predator will be merely to make that individual predator avoid all members of the persecuted species, and so, when the individual attacked possibly survives, to confer no advantage upon its genotype, with gregarious larvae the effect will certainly be to give the increased protection especially to one particular group of larvae, probably brothers and sisters of the individual attacked. The selective potency of the avoidance of brothers will of course be only half as great as if the individual itself was protected: against this is to be set the fact that it applies to the whole of a possibly numerous brood. (Fisher, 1930, p. 178)

Here, Fisher mentions that kin selection directed towards full sibs has only half the strength of individual selection. A full quantitative treatment of kin selection came only with W. D. Hamilton's pair of papers published in 1964, widely regarded as the foundation stone of sociobiology. Hamilton was able to show that the condition for the spread of altruism through kin selection could be succinctly expressed in a simple equation:

$$\frac{b}{c} > \frac{1}{r} \quad \text{or} \quad rb - c > 0,$$

where b is the benefit (in terms of Darwinian individual fitness) that accrues to the beneficiary of the altruism, c is the cost (again in terms of Darwinian individual fitness) that the altruist suffers, and r is the degree of relatedness between the two individuals. r is also the proportion of their genes that two individuals have in common by virtue of **identity of descent**. For example, in an outbred population of a sexually reproducing diploid species the degree of relatedness between a parent and his or her offspring is a half, between an uncle or aunt and a niece or nephew a quarter, and between two first cousins an eighth.

Consider a population of individuals where the only sort of altruism is that individuals invest in their offspring. We now postulate an allele for altruism, A, such that individuals with the genotype Aa are altruistic to relatives other than their offspring, while the aa individuals are the 'selfish' individuals. (This means that we have assumed that A is dominant to a. It can be shown that exactly the same conclusions as reached below result whatever the degree of dominance of A over a – even if A is completely recessive to a.)

Since, in this argument, selfish individuals are aa, when we are considering whether altruism spreads through the population, we have to consider the sort of individuals to whom Aa individuals will be giving altruism, when A is rare. (A separate argument is required once A starts to be common.) So, supposing that A is rare, consider an Aa individual who helps a full brother or sister. Aa individuals will have resulted, ignoring mutations, from matings between Aa and aa individuals. (When A is rare, Aa × Aa matings occur much less often than Aa × aa matings.) Half of Aa's sibs will be aa and half will be Aa. This is why J. B. S. Haldane once remarked that he was prepared to lay down his life for two of his brothers. As only half of Aa's sibs in the above situation will have the allele for altruism, the Aa individual will need to help two of its sibs the same amount as it harms itself. Alternatively it will need to help one sib twice as much as it harms itself.

One crucial thing about the amount of altruism predicted by Hamilton's rule is that it is an **Evolutionarily Stable Strategy (ESS)** in the sense that a population of individuals adopting, or playing (in games theory jargon) the Strategy $rb - c > 0$ cannot be invaded by selfish individuals. Essentially, this is because opportunities arise when altruistic individuals help one another as predicted by kin selection, while the selfish individuals, of course, do not help one another. The result is that the allele for selfishness, a, does slightly less well than the allele for altruism, A.

Subsequent to Hamilton's original work there has been an enormous number of papers and books dealing with the precise quantitative predictions of kin selection theory, and minor modifications of Hamilton's simple rule are now known to apply under certain conditions. However, at worst, Hamilton's formula still provides a reasonably accurate approximation to the conditions under which kin selection operates.

Inclusive fitness

Inclusive fitness was a term introduced by Hamilton intended to simplify the calculation of conditions for the spread of certain alleles by kin selection. Unfortunately, the term has often been misused. Hamilton described inclusive fitness as:

> the animal's production of adult offspring . . . stripped of all components . . . due to the individual's social environment, leaving the fitness he would express if not exposed to any of the harms or benefits of that environment, . . . and augmented by certain fractions of the quantities of the harm and benefit the individual himself causes to the fitnesses of his neighbours. The fractions in question are simply the coefficients of relationship. . . .

Applying this to our earlier example, aa individuals have an inclusive fitness of N, where N is the number of adult offspring they on average raise. The help, b, they sometimes receive from Aa is disregarded as a component of 'the individual's social environment'. The inclusive fitness of the Aa individuals equals $N + (nb)/2 - (nc)$, where n is the number of altruistic acts Aa performs in its life, each at a cost c and of benefit b to a sib ($r = \frac{1}{2}$). A therefore spreads if:

$$N + (nb)/2 - nc > N, \text{ i.e.}$$

$$\frac{b}{2} - c > 0,$$

as before.

Grafen gives examples of erroneous definitions of inclusive fitness, and points out that the reason such mistakes persist is that in a general consideration of kin selection the definition itself is never called on with any precision and so the error does not manifest itself. Once data are used to calculate inclusive fitness, the precise definition obviously does matter.

Kin selection in nature

In a very large number of social species convincing data exist to show that closer relatives receive correspondingly greater aid, and in a few cases Hamilton's predicative equation for the evolution of kin selection has been tested and at least partially verified. A particularly elegant form of kin selection was found in saturniid moths by Blest. Consider two types of insect, one cryptic and palatable, the other aposematic and therefore distasteful, and both subject to predation. The longer that a cryptic insect survives after reproducing, the greater the chances that it will be found by a predator who will learn to recognise and find other individuals of the species. Post-reproductive survival evidently prejudices the survival of the other members of the species including close relatives. The contrary argument applies to the aposematic insect. Predators will learn to avoid such prey. The post-reproductive survival, then, of an aposematic insect should favour the survival of its relatives. Blest's predictions were fulfilled by the data he collected on Barro Colorado Island in the Panama Canal Zone. The aposematic species had post-reproductive lifespans several times those of the cryptic species.

Lions

Lions, *Panthera leo*, have been studied in the Serengeti and Ngorongoro Crater, Tanzania for 20 years. Lions live in stable social groups, called prides, which consist of between 2 and 18 females, their dependent offspring, and a coalition of 1 to 7 adult males. Daughters born into the pride remain there for life while sons leave before they reach reproductive maturity. As a consequence, the females within a pride are quite closely related and **communal suckling** occurs, so that cubs may suck from any adult female with milk.

The adult males associated with a pride may be ousted by other males. As might be expected, large coalitions of males are able to remain with a pride for longer before they are driven away. If a pride is taken over, the new adult males seek out and kill as many as possible of the young cubs. The function of this **infanticide** seems to be to bring the females back into oestrus as, in common with many mammals, female lions do not ovulate while lactating. Infanticide therefore results in the new adult males siring their own cubs more quickly than if no infanticide occurred. It appears that 25 % of all cubs born suffer infanticide (Figure 4.4).

Early data suggested that the adult males in a pride were full or half-brothers. Kin selection was therefore invoked as an explanation for the harmonious relationships between adult males when females are in oestrus. It now appears that 42 % of breeding coalitions of known origins contain non-relatives and that relationships between adult males may not always be so

Figure 4.4 Infanticide by a lion that has found the cub of another male.

harmonious. Fights occur between adult males in which contestants may be blinded. Interestingly, such fights occur just as often between related as between unrelated males.

A remarkable feature of lion reproductive physiology is that during oestrus, which typically lasts about four days, with an inter-oestrous interval of 16 days, copulation occurs on average every 25 minutes. A number of explanations have been suggested for this extraordinary libido. One is that by ensuring that so many copulations are necessary for one fertilisation, females reduce the value to males of each copulations. In consequence, it is not worth the males competing for copulations. This is to the advantage of females, as, with more competition, one or more males in the coalition might be injured. This would be disadvantageous to females since smaller coalitions, as noted above, are associated with more frequent takeovers, and consequently higher levels of infanticide. An alternative and almost diametrically opposed explanation is that heightened sexual activity of females after takeovers attracts other males to their pride range, with the result that prides are taken over again, but this time by larger coalitions, which are again considered desirable. Sociobiologists are rarely short of ideas!

Florida scrub jays

The Florida scrub jay, *Aphelocoma coerulescens*, is a member of the crow family and lives in the south-western states of the USA. Florida scrub jays live in stable groups comprising a breeding pair and their young of the previous one or more years. The groups maintain permanent year-round territories that are defended by all members of the group. In early spring, the breeding pair selects a nest site and builds a nest, unaided; the breeding female alone incubates and is fed on the nest by her mate only. After the young have hatched, however, the non-breeding members provide help in two main ways. First, predation is the major mortality factor facing newly hatched Florida scrub jays, and larger groups of scrub jays are more effective in mobbing and deterring predators (especially snakes) than are single pairs. Secondly, the helpers provide approximately 30 % of the food for the young. Starvation is rare among Florida scrub jay young and the total amount of food brought to nests does not differ between groups with and without helpers. Rather, the effect of auxiliary birds is to enable the parents to reduce their number of foraging trips. Analyses of survivals of breeding scrub jays from one year to the next record a mortality of 13 % among breeders that had helpers, compared to 20 % for breeders that did not. These data demonstrate the important point that one must consider long-term as well as short-term effects whenever calculating the effects of a helper on the lifetime fitness of a breeder.

Data collected by Woolfenden allow a detailed examination of the kin-status of the individual helpers. Of 199 helpers recorded between 1969 and 1977, 118 were assisting both genetic parents, 49 were helping one genetic parent plus a step-parent, and only 32 were aiding other breeding combinations of siblings, grandparents and non-kin.

The tendency to help was age-related. Virtually all one-year-olds were helpers, but the proportion of individuals attempting to breed on their own increased steadily with advancing age. By three and five years of age, respectively, almost all females and males were breeding on their own.

An analysis of the reproductive success of groups with and without helpers clearly demonstrates the positive contribution made by auxiliary birds (Figures 4.5 and 4.6). Evidently helpers help. But do helpers do better to help than to go

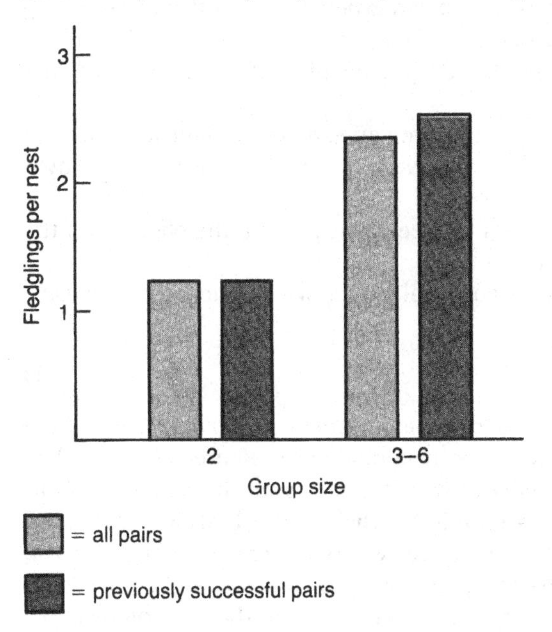

Figure 4.5 Unaided Florida scrub jay pairs (group size = 2) fledge fewer offspring than aided pairs (group size = 3 to 6).

Figure 4.6 (a) Helpers among the Florida scrub jay. At the nest a yearling is assisting the parents feed the nestlings.

off and breed on their own? An attempt to answer this question, using the inclusive fitness approach, has been made by Emlen.

At the beginning of its first potential breeding season, an individual has two alternative **strategies**. It can stay at home and help its parents or it can go off and attempt to rear its own offspring.

Let

N_0 = number of young produced by a novice individual during its first breeding attempt;

N_1 = number of young produced by an experienced member of a breeding pair without the assistance of helpers;

N_2 = number of young produced by a member of an experienced pair aided by helpers;

$\overline{H}$ = average number of helpers aiding an experienced pair with helpers;

r_p = coefficient of relatedness between a breeding bird and its own offspring;

r_h = coefficient of relatedness between a helper and the offspring of the breeding pair it helps.

Then the prediction is that a potential helper will maximise its inclusive fitness only if:

$$\frac{(N_2 - N_1)}{\overline{H}} \times r_h > N_0 \times r_p. \qquad \textbf{(4.1)}$$

For Florida scrub jays, the most difficult parameter to estimate is N_0. There are only 17 occasions known where both members of a pair are novice breeders and no helpers are present. Indeed, by using data from these individuals to estimate N_0, we are implicitly assuming that these novice birds were behaving non-adaptively; that is, that they would have done better to have stayed at the parental nest and helped. Nevertheless, we will proceed as best we can. Estimates for the required parameters are as shown on the table on page 43 (Woolfenden and Fitzpatrick, 1984).

The values of r_p and r_h rely on the observations that the species is outbred – inbreeding was almost never observed – and that helpers usually help their full sibs.

Figure 4.6 (b) Near a Florida scrub jay nest two helpers have spotted an indigo snake, *Drymarchon corais*, a dangerous predator of jay nestlings. One helper crouches on the ground in a threat posture. The other perches nearby in a 'hiccup' stance, which will soon alert the birds at the nest.

Parameter	*Parameter value*	*Number of nests in sample*
N_0	1.24	17
N_1	1.80	64
N_2	2.38	117
$\overline{H}$	1.78	117
r_p	0.50	–
r_h	0.43	–

Putting these figures into equation (4.1) gives us the prediction that helpers should help out if: $0.14 > 0.62$. As 0.14 is manifestly not greater than 0.62 what can we conclude? Perhaps the major conclusion is that without experimental manipulation of communal breeders, the predictions of kin selection will be very hard to test. In the case of Florida scrub jays this is probably because of our difficulties in estimating N_0, as was noted earlier.

Evidently, we can conclude that non-breeding Florida scrub jays should cease to be helpers whenever they can obtain a breeding vacancy. To understand better the situation among scrub jays, it is necessary to examine separately the reproductive options open to non-breeding males and females.

There is a shortage of suitable nesting sites. While remaining on the parental territory a dominance hierarchy develops. The breeding male is dominant, followed by older non-breeding males, younger males, and then females (in turn, age-related). As a group increases in numbers, its territory often expands. This can lead to subdivision or budding off of a portion of the parental territory. This new territory is then occupied by the dominant male helper. Other means, for a male, of obtaining a territory and breeding include occupying vacancies left by the death of nearby male breeders or directly displacing them; the latter is rare in scrub jays.

Females, on the other hand, being subordinate to males, stand virtually no chance of achieving breeding status on their own natal territories. If a son accedes to breeding status on the parental territory he generally mates with a new female who immigrates from outside. Thus a female's strategy lies in surveying different territories, searching for vacancies when a breeding female has succumbed, or finding a new territory. In keeping with these observations, females tend to disperse more often and at an earlier age than males.

Wild turkeys

By contrast with Florida scrub jays and lions, each of which has been studied continuously since the 1960s, our knowledge of the social organisation and behaviour of the wild turkey, *Meleagris gallopavo*, rests on a two-season study (Watts and Stokes, 1971).

Wild turkey females lay a clutch of about 14 eggs in spring and each female incubates her own eggs. The eggs hatch after a month, and the survivors from predators and vagaries of the weather remain together until the autumn when the young are six to seven months old. The young males then leave as a sibling group. This group continues to be an inseparable unit for life. Even if it has been reduced to a single member, the survivor does not try to join another sibling group or form a group with other lone turkey males.

Juvenile male sibling groups aggregate into large winter flocks. It is at this stage in the life of the young male that his **status** is decided. In the exclusively male winter flocks he is forced into two contests: one to establish his position within his own group of siblings, the other to determine the status of his sibling group with respect to other groups. Each sibling engages in physical combat with his brothers. The battle consists of wrestling, spurring, striking with the wings and pecking at the head and neck. Fights often last more than two hours. The strongest fighter in the group becomes the dominant bird, and the order of rank established among the siblings is seldom challenged thereafter as long as the dominant bird lives.

By the end of February, males and females congregate at the four display grounds used in Watts and Stokes' study area. At each display ground there are about 10 to 15 male sibling groups with between them about 30 males. Females visit the area and the male sibling group that has gained dominance over all the other groups moves around with the females, the subordinate groups hanging about at the periphery, taking what opportunities they can to display to females there (Figure 4.7). Displays by male sibling groups usually consist of the members of the dominant sibling group strutting synchronously and close together. Despite this, only the dominant male within the sibling group performs any matings. Indeed, only once in the 59 copulations observed was a mating not achieved by the dominant male of the dominant sibling group at a display area.

The evolutionary question is: why do subordinate male turkeys display synchronously with, and close to, their dominant brothers? The answer proposed by Watts and Stokes (1971) and Wilson (1975) is kin selection. Brother helps brother. An alternative answer is that maybe subordinates are simply making the best of a bad job, practising their strutting in the hope that

Figure 4.7 Courtship in the wild turkey. On the display grounds the brotherhoods, here consisting of two pairs and one solitary male, display to watching females.

they may eventually become dominant. This interpretation receives support from Watts and Stokes' own observation that the average annual mortality of adult male turkeys at their study site is 40 %. It would be helpful to have more data. One can't help wondering – by analogy with Packer and Pusey's studies on lions – whether the members of a sibling group really are always brothers.

Naked mole rats

It is only within the last decade that the most remarkable example so far known of communal breeding within mammals has been discovered. Naked mole rats, *Heterocephalus glaber*, are rodents found in the hot, arid regions of Kenya, Somalia and Ethiopia. They live entirely underground, feeding on roots and tubers, with a deep nest area.

In eusocial insects (see section 4.5 below), a **caste** is any set of individuals of a particular morphological type or age group that performs specialised labour in the colony. By these criteria, naked mole rat colonies have a system of castes. The most detailed observations have come from entire colonies dug up and transported to western zoos where they are allowed to establish themselves in artificial burrow systems (Jarvis, 1981). Perhaps the most interesting finding is that in each colony there is only one breeding female. In one artificially established colony, two females came into oestrus simultaneously and fought violently until one of the pair died.

The remaining mole rats, males and females, are divided into 'workers' and 'non-workers'. Workers dig, forage, transport soil and build nests for the breeding female. Non-workers assist in the care of the young and only very rarely dig or transport materials. Females in both these castes are non-breeding. Histology of their ovaries shows them to have few secondary or tertiary follicles. Spermatogenesis, however, is observed in all adult males.

Young born to the breeding female are kept warm in the communal nest by sleeping mole rats of all castes, but are suckled only by the breeding female. If the colony is disturbed, young are carried from the nest to other parts of the burrow system by workers of either sex. During weaning, the young feed on food brought to the nest by the workers and beg faeces from all individuals except the breeding female. The genetic relatednesses among colony members are not known.

Haplodiploidy

Eusociality is the name given to the condition when a species possesses three characteristics:

1 Co-operation in caring for the young.
2 Reproductive division of labour, with some individuals permanently sterile.
3 Overlap of at least two generations that contribute to colony labour.

Because of the existence of permanently sterile individuals that help other individuals to reproduce, eusociality represents the pinnacle of altruism. It has evolved once in the termites, possibly once in mole rats (as discussed above) and at least eleven times in the order of insects known as the Hymenoptera, which includes ants, bees and wasps.

The Hymenoptera are almost unique in being **haplodiploid**. In haplodiploid species females are diploid but males are haploid and so possess only one set of chromosomes. This genetic peculiarity has very important consequences. Daughters necessarily inherit all of their father's genes, but only pass on half of their genes to their offspring, whether sons or daughters. Outbred females are consequently related to their offspring by $\frac{1}{2}$, but to their sisters by $\frac{3}{4}$ and their brothers by $\frac{1}{4}$. For example, if we consider a female who has one copy of a rare allele, there is a 50 % chance it will be in any of her offspring, a 75 % chance it will be in a diploid sister, but only a 25 % chance it will be in a haploid brother. This asymmetry arises because sisters always share their paternal chromosomes, but a sister never shares her paternal chromosomes with a brother, as he doesn't have any paternal chromosomes! Being haploid, sons develop from unfertilised eggs, produced by **parthenogenesis**.

Haplodiploidy may be the explanation for the prevalence of eusociality in the Hymenoptera. A female ant, bee or wasp should prefer, other things being equal, to produce sisters rather than sons or daughters, as she is more closely related to her sisters than to her offspring. This agrees well with the observation that in all the Hymenoptera workers are only females, whereas in termites and mole rats workers may belong to either sex.

Humans

All of us know that kinship is important in human relationships. One step forward taken by sociobiology is to make quantitative predictions about the extent to which kin selection favours relatives. Sadly, however, the required data are difficult to collect from wasps and Florida scrub jays, let alone from humans. Nevertheless, it is natural to ask whether kin selection is responsible for at least part of the capacity for altruism in humans. This suggestion gains some support from the probable fact that during most of our hominid history the predominant social unit was the immediate family and a tight network of other close relatives.

In a few societies, fathers take little or no interest in their wife's children but direct their paternal care towards their sisters' children. As a sister's children are less closely related to a father than his own offspring, this practice appears paradoxical. Consider, however, what happens when there is a good deal of uncertainty about paternity. Sisters may only be half-sisters, but they and their offspring are indubitably true kin. On the other hand, a wife's children may often be unrelated to the husband, if adultery is rife. A review of the literature provides some support for the prediction that it is in precisely those societies in which there is greatest uncertainty about paternity that men divert a substantial part of their paternal care to their nephews and nieces.

Humans are perhaps unique in that the female of the species exhibits a menopause. It has been suggested that postmenopausal women may exhibit kin-selected altruism. The idea is that, freed from their own reproductive constraints, such women have the time and energy to help in the care and rearing of their other relatives including, for example, their grandchildren.

Reciprocal altruism

A second way, besides kin selection, in which altruism can evolve is through **reciprocal altruism**. Essentially, reciprocal altruism is: 'You scratch my back, I'll scratch yours'. Perhaps the most convincing example occurs in male olive baboons, *Papio anubis*. Craig Packer studied 18 adult males in 3 troops at Gombe National Park, Tanzania, for more than 1100 hours. All males leave their natal troops and transfer to another troop before reproducing. This means that troops contain several males unrelated to each other. Olive baboons form exclusive consort pairs consisting of an adult male and an oestrous female. These consort pairs may last for several days. Coalitions between unrelated males are sometimes formed in attempts to separate an opponent from an oestrous female. If a pair of males does succeed in obtaining a female in this manner, only one of the two coalition partners mates with her. Attempts at enlisting a coalition partner can be recognised unambiguously: one individual, the enlisting male, repeatedly and rapidly turns his head from a second male, the solicited individual, towards a third male, the opponent, while continuously threatening the third.

Packer saw 20 occasions on which coalitions formed when the opponent was consorting with an oestrous female. In six of these cases the formation of a coalition directed against the consorting male resulted in the loss of the female by the single opponent. In all six cases the female ended up with the enlisting male of the coalition; the solicited male generally continued to fight the opponent while the enlisting male gained access to the oestrous female. In each case the solicited male evidently risked injury from fighting the opponent, while the enlisting male gained access to an oestrous female.

The crucial fact that Packer discovered was that males sometimes reciprocated in joining coalitions at each other's request. Individual males that most frequently gave aid were those that most frequently received aid. Furthermore, males had preferences for particular coalition partners based at least partly on reciprocation: for nine of the ten males who solicited other males on four or more occasions, the favourite partner in turn solicited the original more often than the average number of occasions that the partner solicited all adult males in their troop.

Empirical evidence of reciprocal altruism also comes from chimpanzees and vervet monkeys, *Cercopithecus aethiops*. As Trivers realised, a fundamental difficulty in the evolution of reciprocal altruism is the possibility of **cheating**:

'You scratch my back, then I run away'. The preconditions for the evolution of reciprocal altruism are similar to those ideal for kin selection: long lifetime, low dispersal rate, and mutual dependence. These make it difficult in any one case to distinguish between the alternatives of kin selection and reciprocal altruism.

While kin selection evidently operates in humans, reciprocal altruism is probably of particular importance. Reciprocation among individuals, related or not, is the key to much of human society. As Wilson (1978) writes:

> The 'altruist' expects reciprocation from society for himself or his closest relatives. His good behavior is calculated, often in a wholly conscious way, and his maneuvers are orchestrated by the excruciatingly intricate sanctions and demands of society. The capacity for [reciprocal] altruism can be expected to have evolved primarily by selection of individuals and to be deeply influenced by the vagaries of cultural evolution. Its psychological vehicles are lying, pretense and deceit, including self-deceit, because the actor is most convincing who believes that his performance is real.

Group selection

The third way in which altruism can theoretically evolve is by **group selection**. Wynne-Edwards, primarily concerned with attempting to explain how population density is regulated, argued that populations should not overexploit their food supplies, for such overexploitation would lead to reduced food yields and thus to lower reproductive success. As an example, he described how the potential yield of many fisheries was drastically reduced when they were overfished; but could recover when catches were voluntarily restricted. Similar processes should apply, he argued, to non-human population, which should restrict their population density and rate of reproduction rather than endanger their food supply.

As Wynne-Edwards' theory depends on the assumption that individuals do not always maximise their own reproductive success, it was necessary to extend evolutionary theory to account for this. Wynne-Edwards argued that groups containing individuals who reproduced too fast, so that the recruitment rate persistently tended to exceed the death rate, must have repeatedly exterminated themselves by overtaxing and progressively destroying their food sources. Prudent groups, where altruistic individuals restrained their reproduction, would outlive more selfish groups, and so come to predominate.

Most theoreticians, with a few notable exceptions, have accorded Wynne-Edwards' theory of group selection a poor welcome. Group selection faces a fundamental theoretical problem. Imagine a prudent group whose members are reproducing submaximally due to group selection. Suppose now that a mutation arises which causes its holder to reproduce maximally. Even if such a group now becomes more prone to extinction as group selection requires, the altruistic individuals within it would, by definition, produce fewer surviving descendants than the other group members bearing the new mutation. Consequently, selection acting within the group would tend to eradicate the altruistic trait. Group selection only becomes important when two conditions are fulfilled: first, that selfish groups become extinct much more quickly than prudent groups, and secondly, that little migration takes place between groups. When migration becomes frequent, selfish individuals tend to move between groups before groups become extinct. The available data on animal migration and demography suggest that group selection, certainly in vertebrates, is unlikely to be important.

The selfish gene

It will have become apparent just how much discussion of animal behaviour has become dominated by thinking at the level, not of groups or even individuals, but of genes. Essentially, this is because genes last longer than groups or individuals (Dawkins, 1976). This gene's-eye view of behaviour has even reached the point where Richard Dawkins felt compelled to include in a more recent book a chapter titled 'Rediscovering the Organism', which addresses the problem of why genes clump into cells and cells into the multicellular clones we call individuals.

4.5 The social life of insects

As Wilson (1975) notes, the social insects challenge the mind by the sheer magnitude of their numbers and variety. There are more species of ants in a square kilometre of Brazilian forest than there are species of primates in the entire world; more workers in a single colony of driver ants than all the elephants, lions and tigers in both Africa and India.

The degree and extent of insect sociality is tremendously varied, yet the subject is in its infancy. The great majority of species are totally unknown behaviourally. The best-studied group of social insects is the ants, yet probably fewer than two-thirds of the extant species have ever been assigned a name in the binomial system. Of the perhaps 12 000 living ant species only about 100 have been studied with any care.

In the face of such ignorance, and much evident behavioural diversity, generalisations about insect sociality should be made warily. We have already defined eusociality, and considered the importance of haplodiploidy in its evolution (section 4.4). Some general comments can be made, and then just one social species – the honeybee – will be considered.

Once a species has crossed the threshold of eusociality, there are two complementary means by which it can advance in colony organisation; first, through an increase in the numbers of the worker castes, and their degree of specialisation, and secondly, through the enlargement of the communication code by which the colony members co-ordinate their activities. Communication in the honey bee is reviewed in Chapter 6. Oster and Wilson defend the theory that castes tend to proliferate in evolution until there is one for each task. This is not as incredible as it may at first sound. In a series of studies on ecologically disparate species in five ant genera, the total behavioural repertoire of the worker caste has been estimated to lie between 20 and 45. The number of castes, it is argued, should equal the number of tasks to maximise colony efficiency, in the same way that in an assembly line each worker performs only one operation. In nature, the number of castes almost never equals the number of tasks, even when **temporal castes** as well as **morphological castes** are taken into account. An individual may pass through several temporal castes as she develops. Within the Hymenoptera no certain case of morphological or temporal caste polymorphism has so far been found in males.

Other evolutionary trends can be recognised. As colony size has grown in the course of evolution, the differences between the queen and worker castes have been exaggerated, until queens ultimately serve only to lay eggs. Correlated with this trend has been a subtle shift in the power structure of the colony. Among primitive social insects, the queen maintains a dominant position primarily by aggressive behaviour. In the most complicated social species, control is exercised through inhibitory pheromones.

Honeybees

The common honeybee, *Apis mellifera*, can be taken as representative of the most advanced social bees. By the general intuitive criteria of social complexity – colony size, magnitude of queen–worker difference, altruistic behaviour among colony members, periodicity of male production, complexity of chemical communication and regulation of the insect temperature – the honeybee has few peers (Figure 4.8). In one feature, the waggle dance (see Chapter 6), the species perhaps stands truly apart from all other insects.

There is still a tremendous amount we do not know about honeybees. For example, it used to be believed that each queen mated only once in her life, but we now know that most queens mate with several drones, often in quick succession on the same nuptial flight, before laying any eggs. This observation

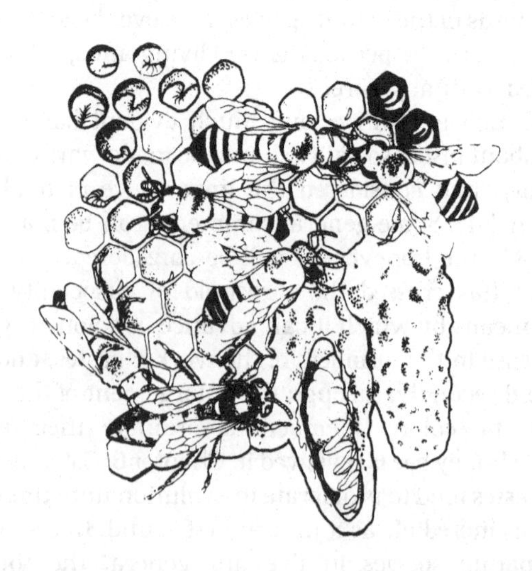

Figure 4.8 A portion of a honeybee colony. In the lower lefthand corner the queen is surrounded by a typical attendant. She is resting on a group of capped cells, each of which encloses a developing worker pupa. Many of the open cells contain eggs and larvae in various stages of development, while others are partly filled by pollen masses or honey (extreme upper right). In the upper right a worker extrudes her tongue to sip regurgitated nectar from a sister. At the lower margin of the comb are two royal cells, one of which has been cut open to reveal the queen pupa inside.

invalidates Hamilton's explanation for the importance of haplodiploidy in the evolution of eusociality discussed earlier (section 4.4 above). Once **multiple insemination** becomes common, females are not necessarily more closely related to their sisters than to their offspring.

After mating, the queen returns to her hive and is fed on 'brood-food', a proteinaceous and highly nutritious substance secreted by the brood-food (pharyngeal) glands of workers. The size of the queen increases – though nothing like as much as in termite queens – and within three or four days she begins egg-laying. At the peak of its numbers in summer, a healthy colony of honeybees consists of up to sixty thousand **workers**, a few hundred males (**drones**) and a single fertilised **queen**, together with a number of combs containing **eggs**, **larvae** and **pupae** as well as stores of honey and pollen.

Division of labour among the workers has been known since the time of Aristotle. Most workers live about six weeks, though those that emerge in autumn may live four times as long. Workers exist as temporal castes. The youngest workers either clean out brood cells from which bees have recently emerged, or remain motionless on the brood combs helping to incubate the brood. By the time a worker is five or six days old her brood-food glands begin to secrete and she feeds the younger larvae on brood-food. This she continues to do until she is ten or twelve days old by which time her brood-food glands have become greatly reduced in size. However, by now the wax-producing glands on the underside of her abdomen have gradually become active and by the time she is about twelve days old, she is ready to begin comb-building and repairing, which requires the production of fresh beeswax. When a worker is about three weeks old she takes up the duty of foraging for water, nectar or pollen, and this she proceeds to do until she dies of old age one or two weeks later. It is usually some of the younger foragers who guard the entrance of their hive or nest against intruders; but most workers omit this duty altogether, otherwise there would be too many guard bees on duty. Such temporal castes are reminiscent of the weekly rota of duties of some Victorian maids. Presented with a clean apron on Monday (at 5a.m.) they would be in charge of making the beds. On Tuesday they might help with clothes washing. By the time Saturday arrived they would be cleaning the fire grates.

What it is that causes some diploid cells to develop into workers and others into queens is still not fully known. No one has yet succeeded in rearing queens in the laboratory. Queen cells are larger than worker cells and hang almost vertically, whereas worker cells lie horizontally. For the first three days, queen larvae and worker larvae appear to be treated identically. After that, however, only the queen larvae continue to be fed on 'royal jelly', a highly nutritious food secreted by the hypopharyngeal and mandibular glands of the honeybees. Worker larvae are switched from this to a mixture including pollen and nectar (or dilute honey).

4.6 The social life of apes

The apes are one family in the primates and include only the gibbon, the orangutan, the chimpanzee and the gorilla. Of these four, the two that have received the most intensive study are the chimpanzee and the gorilla.

Gorillas

There are two subspecies of gorilla, the lowland gorilla, and the one about which much more is known, the mountain gorilla, *Gorilla gorilla beringei*. Until Schaller spent a year and a half watching them, almost nothing was known about their behaviour. Subsequent to Schaller's investigation, Dian Fossey spent 20 years studying them, until she was murdered by poachers in the Christmas of 1985.

Gorillas are the largest of the primates. The mountain subspecies thrives in a diversity of habitats from lowland rain forest to the thick bamboo stands and *Lobelia–Senecio* groves of the high mountains. The common denominator is a preference for humid environments and low, verdant vegetation, as gorillas, being vegetarians, spend most of their waking lives eating.

Gorillas live in long-lived groups of 2 to 30 individuals. A typical group might have a single dominant silver-backed male, a couple of younger adult black-backed males, half a dozen females and a comparable number of immature individuals. Lone males and occasional bachelor groups also occur. The gorilla troops occupy a home range that changes only slightly over a period of weeks. Home ranges overlap extensively and territorial defence is absent. Nevertheless, spacing occurs in the sense that the centres of the home ranges are spaced out fairly regularly rather than randomly. Encounters between neighbouring groups are usually peaceful; the groups continue to feed or progress in full view of each other without visible excitement and sometimes even mingle for a few minutes. But mutual aggression, consisting of charging, also occurs on occasions.

Loud hoot-calls are emitted only by silver-backed males and only during exchanges with other groups or solitary males. The distance between two silver-backs calling may be as little as six metres or as much as a kilometre. In spite of the general slow tempo of gorilla social life, the cluster of individuals seldom exceeds 70 metres in diameter, and the dominant male is always within easy vocal range of the other troop members.

Big-game hunters had described gorillas as ferocious. Schaller's and Fossey's studies have shown that silver-backs may protect their troop from other males or from poachers, but that otherwise gorillas are shy and gentle. Indeed, in her first 3000 hours of observation, Fossey saw just 5 minutes of hostility! More recent research by her, however, has shown the importance of long-term research in our understanding of animal behaviour, particularly when the species one is investigating may live 60 years. As with lions, and for the same reason, infanticide occurs and is now known to have accounted for the deaths of 6 of the 38 infants born during the study. It is not only infants who may be killed. Since Fossey's study began in 1967 she had observed a silver-back called Beethoven, believed to have been born in the late 1920s. Beethoven had a vigorous ally in his son Icarus. When he was 11 years old Icarus attained sexual maturity and the relationship between the two males became less harmonious. As Beethoven weakened with age, antagonism intensified between father and son on the days in each month when the adult females were in oestrus. Harsh and agitated vocalisations created a great deal of friction within the group. The conflict reached a dramatic climax when Icarus murdered Marchessa, one of the oldest females in the troop. Beethoven was devastated by the death of Marchessa. Often he was heard whimpering, a sound never before recorded from silver-backs. Now Icarus commands the entire troop, though Beethoven is still a member, and Icarus can mate with any adult female other than his mother, Effie.

Chimpanzees

When Jane Goodall first arrived in 1960 to study the chimpanzees at the Gombe reserve on the banks of Lake Tanganyika, the game warden who took her round made a mental note that she wouldn't last more than six weeks. She has stayed for over 20 years, producing the classic account of chimpanzee social organisation and behaviour in her fascinating and moving book, *In the shadow of man* (van Lawick-Goodall, 1971).

By comparison with gorillas, chimpanzees (*Pan troglodytes*) are organised into larger societies, within which casual groups form, break up and re-form with extraordinary fluidity. Chimpanzees occur widely in forested areas throughout equatorial Africa. They are semi-terrestrial, spending 25–50 % of their time on the ground. They forage during the day and build sleeping nests, as do gorillas, in trees for the night. Unlike gorillas, chimpanzees are omnivorous, feeding to a large extent on fruit but also on the leaves, bark and seeds of a wider variety of plant species. They collect termites and ants and occasionally kill and eat small baboons and other monkeys.

A high level of co-operation occurs between chimpanzees. Most of the time members of a party feed separately on fruit and other vegetable items. But, if the supply is limited, chimps beg from one another. Co-operation, usually between males, is also apparent when small animals are hunted for food. Chimpanzees are unusual among mammals, though not unique, in that it is the females who at puberty disperse from the troops in which they were born. This means that in a group, males may be highly related, accounting at least in part for their co-operation in hunting and frequent grooming bouts. It also means that mothers may develop much closer links with their sons than in other species.

In 1964, when her daughter, Fifi, was about six years old, Flo gave birth to Flint. Flo was an excellent mother, affectionate, tolerant and playful – and also high-ranking. Flint grew up normally, but when Flo, in an effort to wean him, tried to prevent him from taking milk, he would throw tantrums. He even hit and bit his mother, behaviour rare in youngsters. Flo, often seeming to lack the energy to cope with Flint, would give in. In 1968 Flo gave birth to a female infant. Flo's problems with Flint increased. Youngsters normally become quite independent after the birth of a sibling, but Flint persisted in riding on Flo's back, despite the new baby clinging below, and he insisted on pushing into the family nest at night. When the baby was six months old, Flo became very ill. By the time she eventually recovered, the baby had disappeared and she was prepared to accept Flint's infantile behaviour. In some ways he seemed to fill the place of her lost baby. Four years later Flo died of old age. Flint, then aged eight, had still been sharing a nest with her at night. Flint immediately became lethargic and depressed. He scarcely ate and seldom interacted with other chimpanzees. Three weeks after Flo's death, Flint himself died.

Jane Goodall's more recent research has shown that, as with gorillas, chimps have what seems to us a less pleasant side of their nature. In 1970 her main study community divided into a southern Kahama group and a northern Kasakela group. In 1974 a gang of five chimpanzees from the Kasakela community caught a single male of the Kahama group. They hit, kicked and bit him for 20 minutes and left him bleeding from innumerable wounds. He died. A month later another prime Kahama male was caught by three males from Kasakela and severely beaten up. He became terribly emaciated, with a deep unhealed gash in his thigh, and died. Goliath, an old male, was next, then an old female, Madame Bee. Eventually, the entire Kahama group was exterminated. At the same time, one of the adult females in the Kasakela community took to killing and eating infants, with the help of her daughter. Between them they are known to have killed and eaten at least three infants.

Researchers studying the higher primates seem almost always to consider their study animals as individuals. As one reads the detailed studies of primates, one notices that animals are known as individuals and are named. Differences between individuals in their behaviour are considerable and enrich descriptions of the species' social organisation. Such inter-individual differences point towards the full diversity of human society considered in Chapters 7 and 8.

5 Reproductive behaviour

Animals have a limited amount of time and energy that they can devote to reproduction. During an individual's life he or she will make a number of 'choices' about when to mate, with whom to mate, how many resources to devote to each offspring, and so on. These 'choices' may or may not be conscious ones. To avoid longwindedness, evolutionary biologists tend to talk about individuals making 'choices' when often they mean context-specific behavioural strategies fixed by natural selection, perhaps millions of years ago. Of course, in some organisms, particularly the higher primates including humans, some decisions may truly result from a conscious weighing of the arguments for and against a particular course of action. Even in humans, however, much reproductive behaviour is an innate product of natural selection, while an individual's course of action must be determined in part by the expectations of others in his or her society.

5.1 What sex to be

In most species, of course, an individual's sex is fixed at conception by the inheritance of sex chromosomes or by some other genetic mechanism. Even among vertebrates, however, there exist species in which individuals can change their sex during their lifetime. In anemonefish of the genus *Amphiprion*, for example, sexually immature individuals, if they do become sexually mature, always develop into functional males. Subsequently, they may change into females. The reason for this becomes apparent once the social organisation of the anemonefish is examined. In this genus the typical social unit consists of an adult female, an adult male and several unrelated sexually immature juveniles inhabiting a single sea anemone. Adults are fairly sedentary. In one investigation, some individuals were observed in the same sea anemone over a period of three years.

In *Amphiprion*, the larger of the adult pair is a female. The male behaviourally dominates the juveniles, thus preventing them from developing into competing males. These non-breeding individuals are most likely tolerated because they help defend the anemone and provide replacements if either of the adult pair dies. If the adult female dies, the adult male changes sex, becoming a female, and one of the juveniles becomes an adult male. If the adult male dies, he is simply replaced by one of the juveniles. Selection appears to favour a breeding group size of two. With just two fertile individuals, the larger should be a female, as the production of fertilised eggs by the pair is limited more by the egg production rate than by the sperm production rate. Consequently, all the fish on the sea anemone should want the largest individual to be a female. The adult male might gain if the smaller individuals were also egg producers, but they appear in some way to be prevented from so being by the dominant female.

In the same family as anemonefish (the Pomacentridae) are damselfish of the genus *Dascyllus*. Here, the breeding group consists of three or more fertile adults. In this genus the largest individual is a male, and he can fertilise the two or more adult females whose combined egg output is presumably greater than his potential egg production as an adult female. In damselfish, sexually immature individuals, if they do mature sexually, first develop into adult females. Subsequently they may change into adult males.

5.2 Choosing a mate

In many species, **inbreeding** – breeding with close relatives – leads to **inbreeding depression**. Fertility is reduced and offspring are more likely to die and suffer congenital deformities. Data from humans are understandably rare, but convincing in the case of extreme inbreeding. Matings between fathers and daughters, brothers and sisters, and mothers and sons produce offspring with a 6–20 % increase in death or major defect by comparison with an outbred control group.

Evidently, such inbreeding in humans is biologically disadvantageous, both to the parents – particularly the mother who will inevitably have invested considerable energetic resources in the offspring once conceived – and to the resultant son or daughter. How then is inbreeding avoided? One way is through **incest taboos**, in which certain categories of near relatives are forbidden as marriage partners. However, marriage taboos do not accord perfectly with the biological predictions of inbreeding avoidance. In the majority of the marriage-types forbidden by the Church of England, the parties are unrelated, for example husband and wife's sister.

In addition to marriage taboos there is a second way in which inbreeding is avoided. At least regarding brother–sister incest in humans, there is good reason to believe that inbreeding would be unlikely even in the absence of culturally imposed taboos. The evidence for this comes from children raised like brothers and sisters, but between whom no barriers of biological kinship or incest taboos exist. Studies of young adults raised on Israeli kibbutzim have revealed the remarkable fact that they never marry within their rearing groups. The critical factor is very clearly that the children are raised together from an early age, since the only exceptions involved pairs who had been separated from one another for a large part of their childhood. The records of thousands of married adults reveal not a single marriage between a man and a woman raised together throughout childhood.

If rearing boys and girls together prevents mutual sexual interest, it should normally function as an effective outbreeding mechanism. In most circumstances, including those in which most hominid evolution presumably took place, children reared in very close proximity are likely often to be kin, and the mechanism will function to ensure outbreeding. Interestingly, in the practice of sim-pua marriage in Taiwan, in which the couple are married as children and reared together, such couples have more difficulty consummating their marriages, more extramarital affairs, more divorces and 30 % fewer children than do couples whose marriages are arranged post-pubertally.

In the face of the above data, the existence of first-cousin marriages at first seems surprising. Such marriages are, however, widespread. For example, the Dravidian peoples of the four South Indian states of Andhra Pradesh, Karnataka, Kerala and Tamil Nadu numbered 164 million at the 1981 census. For more than 2000 years they have demonstrated a preference for inbred marriages, averaging in excess of 30 % consanguinity per generation. (By comparison, a marriage between a half-brother and a half-sister, or a grandfather and a granddaughter, in an outbred population represents 25 % consanguinity.) Why does such inbreeding occur? The traditional anthropological answer is that marriages function not so much to maximise the number of children surviving (i.e. the reproductive success of the parents) but to cement political, economic and family ties. A more recent biological answer is that too much outbreeding may itself be a bad thing! The best experimental evidence for this so far comes from a study of the plant *Delphinium nelsoni*. Here, fitness declines with increasing geographical distance between conspecific parents within populations, an effect perhaps caused by the destruction through recombination of local adaptive gene complexes. Interestingly, Bateson has found that Japanese quail preferred to mate with first cousins rather than siblings, third cousins and unrelated conspecifics. He controlled for familiarity by rearing his experimental quail in isolation from each other.

The choice of a mate involves much more than choosing an individual of the right degree of relatedness. At the most basic level a mate must be an individual of the same species but opposite sex. As females generally invest more in their offspring than do males, **mate choice** is particularly important for females. Accordingly, males are selected to make themselves attractive to females, displaying their fitness-enhancing characteristics of good genes, adaptive behaviour, or valuable resources. **Courtship** often provides opportunities for would-be mates to prove their competence as provisioners, and females in particular are likely to mate preferentially with males that provide reproductively relevant resources, either directly or by virtue of the quality of their territories. In the common tern, *Sterna hirundo*, for example, the greater the number of fish the male brings to the female during courtship, the more likely the female is to mate with him, and the more offspring subsequently fledge. For their part, males often prefer to mate with larger, and consequently more fecund, females. In certain cases, males deprive females of any choice by raping them. For most duck species the number of rapes increases as the male parental role decreases, possibly because the absence of a paternal male makes single females more susceptible. In mallard, males respond to rape of their mates by forcing a copulation themselves, introducing their sperm to compete as quickly as possible with that of the rapist.

Many humans, of course, never marry and never have children. Abstention from sexual intercourse occurs in many other animals, too. In many species reproduction may be limited, for males in particular, to the larger or older individuals. Among social insects, countless individuals belong to sterile worker castes. Can cultural phenomena, such as celibacy in humans, be interpreted within a biological framework? It has been suggested that where environmental circumstances are such that life is difficult and the chances of survival relatively low, religions tend to emphasise high levels of reproduction and are generally pro-reproductive in tone, approving of marriage and the rearing of large families. This is the case, for example, with Islam, which has a large number of pro-reproductive characteristics and does not advocate celibacy. Islam is a religion which became established and has thrived principally in the arid regions of the Middle East and parts of Africa. In countries like Tibet, Ladakh and Thailand, the number of celibate monks may constitute up to a third of the total adult male population. In such cases the effect on the birth rate is by no means negligible, and it has been argued that celibacy has evolved as a population control device. It is difficult, however, to see how celibacy could evolve in this way by natural selection, as group selection is implicated. As suggested earlier (section 4.4) group selection is unlikely to be an important force in evolution.

One possibility is that monks, who possess no land, are encouraged by families because they increase the chance that estates will remain undivided. Typically, the eldest son inherits the estate, while monks are usually second sons. Another possibility is that celibacy is indeed a population control device, but that it exists despite natural selection. Yet another possibility is that in some human societies migration rates, other demographic variables and ecological circumstances are such that group selection can be favoured by natural selection. It is not easy to untangle the biological and cultural bases for human sexual and other social behaviour.

In homosexuality a 'mate' is chosen for what appear definitely not to be biological reasons. While the incidence of overt human homosexuality evidently varies between cultures, homosexual behaviour of one form or another is common in virtually all cultures and has been permitted or approved in a number of societies: in classical Athenian, Persian and Islamic societies, in late republican and early imperial Rome, in the urban, Hellenistic cultures of the Middle East, in the Ottoman Empire, and in feudal and early modern Japan. Kinsey's classic studies in the USA concluded that as many as 2 % of American women and 4 % of American men were exclusively homosexual, while 13 % of men were predominantly homosexual for at least three years of their lives.

It has been suggested that homosexuality is normal in a biological sense, and that it is associated with altruism that evolved as an important element of early human social behaviour. Studies of twins suggest that homosexuality may have some genetic basis. If this is the case, the existence of such homosexuality presents an intellectual problem. How would genes predisposing their carriers towards homosexuality exist in a population if homosexuals have fewer than the average number of children? One possibility is that homosexuals, freed from parental obligations, would have been in a position to operate with special efficiency in assisting close relatives. They might further have taken the roles of seers, artists, and keepers of tribal knowledge. If the relatives of homosexuals benefited thus, the genes these individuals shared with the homosexual relatives would have increased at the expense of alternative genes. Inevitably, some of these genes would have been those that predisposed individuals towards homosexuality. A proportion of the population would consequently always have the potential for developing homosexuality. It must be emphasised that such a theory for the evolution of homosexuality is highly speculative.

5.3 Sons or daughters?

Once mating has occurred, should parents produce sons or daughters? In a large number of haplodiploid species it is known that adult females can adaptively choose whether to produce sons or daughters. In parasitic haplodiploid wasps, fertilised eggs may be laid in large hosts, unfertilised eggs in small hosts. This is functional because large daughters go on to enjoy greater reproductive success than large sons, while small daughters produce fewer offspring than small sons. In haplodiploid species, mothers can determine the sex of their offspring simply by controlling whether sperm have access to eggs. Perhaps it is not surprising that the evidence for adaptive parental control of offspring sex ratio is considerably poorer in diploid species. Two explanations have been suggested for the fact that, in diploids, adaptive variation in progeny sex ratio is rarely if ever seen. One is that parents are simply unable to control the sex ratio at conception, while subsequent alterations (e.g. by sex-specific abortion) result in too costly a loss of zygotes. The other is that genes in the sperm actively oppose such parental control of the sex ratio.

In humans it appears that parents will in some cases practise sex-specific infanticide on their own offspring. According to a demographic survey by Divale of over 600 'primitive' societies, there is a remarkable imbalance in the ratio of boys to girls, with an average of 150:100. Divale argues that many societies follow a practice of overt female infanticide. Daughters are suffocated or simply left unattended away from home. But more often the infanticide is more subtle – the discrepancy between male and female infant mortality rates usually results from a neglect of daughters, rather than from a direct assault on the baby's life. Even a slight sex-difference in a mother's responsiveness to her children's cries for food or protection might cumulatively lead to imbalance in adolescent sex ratios.

Why, though, are daughters more likely to be killed than sons? At least part of the answer is that males are more likely, both through war and disease, to die from other causes than are females. Sons are consequently, on average, more valuable than daughters. The fact that in humans the most reproductively successful males sire far more descendants than the most reproductively successful females means that in most societies the killing of daughters should be particularly pronounced in the upper classes, as parents in these classes will produce more descendants through sons than through surviving daughters.

In pre-colonial and British India, the upward flow of daughters by marriage to higher ranking men (termed hypergamy) was sanctified by custom and religion, while female infanticide was practised routinely by the upper castes. After all, in such a scheme of things, whom are the females of the upper castes to marry? The Bedi Sikhs, the highest ranking priestly subcaste of the Punjab, were known as Kuri-Mar, the daughter-slayers. They destroyed virtually all daughters and invested everything in raising sons who would marry women from lower castes. In pre-revolutionary China, female infanticide was commonly practised by many of the social classes, with essentially the same effect as in India – a socially upward flow of women accompanied by dowries, a concentration of both wealth and women in the hands of the middle and upper castes, and near exclusion of the poorest males from the breeding system.

Human infanticide need not be sex-specific. In many societies there is a birth-spacing mechanism in which ovulation in a mother is inhibited during the period of lactation. Such a birth-spacing mechanism is often reinforced by taboos against sexual intercourse for some time after the birth of a child. The duration of the taboo period varies considerably, as does the faithfulness with which it is observed. The most common duration is about six months. In practice, the mechanisms of birth spacing do not guarantee complete avoidance of competition between successive infants detrimental to the survival prospects of one or both. Two dependent infants can be a severe strain on the resources of a mother. Many people the world round routinely sacrifice one of a pair of twins unless unusually favourable circumstances suggest that both can be reared, and when single births occur too close together, mothers often resort to infanticide. One mother in the South American tribe of the Yanomamö told an anthropologist studying her tribe that she had been obliged to kill her latest baby because it would have taken milk away from her unweaned son. Careful demographic surveys demonstrate statistically that the older a first child is at the birth of a second, the more chance the second has of surviving.

5.4 How much to invest?

Once offspring, of whatever sex, have been produced, parents, or occasionally other individuals, have to allocate a certain amount of **parental investment** to them. The question of how much investment should be made in an offspring depends on the gender of the parent. In some species females do most of the investing (e.g. most mammals); in some species males invest more than females (e.g. sticklebacks); in some species both sexes invest a lot in each offspring (e.g. many birds); while in other species each sex invests only a little in each offspring and as many offspring as possible are produced (e.g. many fish and most invertebrates).

Games theory analysis suggests that there is often a simple explanation for which sex invests more: it's the one left holding the baby, literally. Consider a species with internal fertilisation. Here, males can fertilise the eggs and depart immediately, leaving females to bring up the young. With external fertilisation, however, the tables are reversed and males are predicted to be the sex investing more, as females can depart once the eggs have been laid and while the male is ejecting his sperm. These predictions seem often to be borne out, though exceptions are known.

It is easy to assume that reproduction is in each offspring's interest. Sociobiology claims this is an oversimplification. In sexually outbreeding species, parents and offspring are not genetically identical. Then there is a conflict of interest between them. At some stage in a parents' investment in an offspring, the parents have to decide to stop investing in its present offspring A, and start investing in another offspring, B.

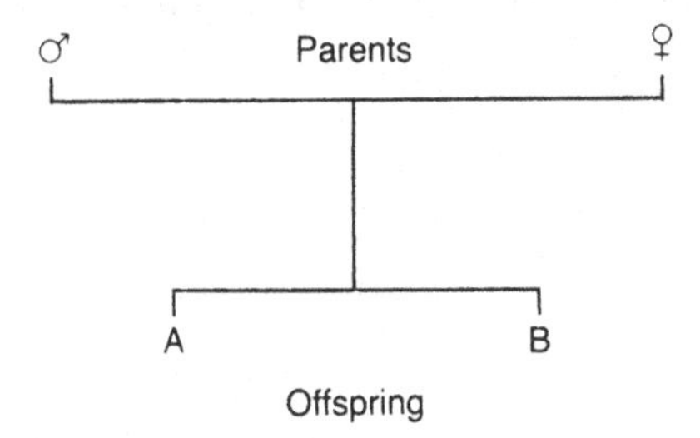

Now each parent is related equally to A and B. A, however, is more closely related to itself ($r = 1$) than it is to B ($r = \frac{1}{2}$ for full sibs in a diploid species; $r = \frac{1}{4}$ for half-sibs). Consequently, while the parents have the interests of A and B equally at heart, A is more interested in itself than in B.

How is this conflict of interest likely to manifest itself? It is expected that each offspring should want more parental investment than the investing parents are prepared to give. Trivers suggested that this might be the explanation for weaning conflict in mammals, as such conflict occurs when the offspring wants more parental investment than the parent is prepared to give. A species where this is predicted not to be the case is the nine-banded armadillo. In this species, young are always born in litters of identical quadruplets. Consequently, in a given litter, there should be no sib–sib competition for access to parental resources, or any other for that matter. (There should, however, be between-litter conflict.) This prediction has been around since at least 1976 but as far as we know nobody has undertaken the required study.

Blurton Jones has pointed out that **parent–offspring conflict** might be the explanation for the fact that in humans, offspring birth weight is on average suboptimal in the sense that larger-than-average offspring generally survive better. Heavier babies would, of course, require more parental investment. There is, however, an alternative explanation. Larger than average babies might result in complications at birth due to too narrow a maternal pelvis, leading, in societies with only primitive obstetrics, to the death of mother and child.

The optimal amount for a parent to invest at any one moment may depend on the sex of the offspring. For example, in red deer, *Cervus elaphus*, male calves have a longer gestation length, weigh more at birth and are suckled more often and for longer. However, sons disperse from their mother's home range area when aged between two and four years, whereas daughters remain in their natal area all their life. Hinds (adult females) belonging to large matrilineal groups show reduced reproductive success, which suggests that mothers continue to invest in their daughters beyond weaning, by allowing their daughters to share their home range area, and presumably, therefore, their food (Clutton-Brock, Guinness and Albon, 1982). Selection has probably favoured hinds which invest heavily in their sons before weaning, not least because sons who show the greatest juvenile growth rates go on to become particularly large, and so achieve high reproductive success through their superior strength in fights for harems. Tim Clutton-Brock suggests that, overall, red deer investment in sons equals investment in daughters (Figure 5.1) as predicted by evolutionary theory.

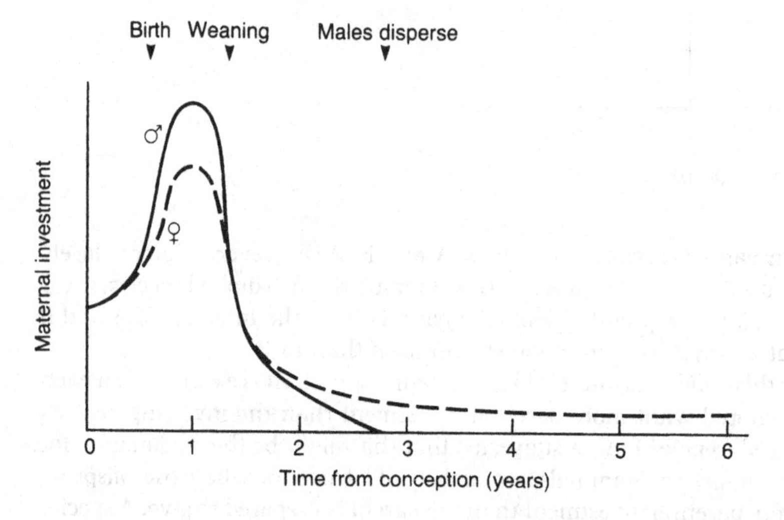

Figure 5.1 Possible pattern of maternal investment in red deer in daughters (♀) versus sons (♂).

6 Communication

Many of the most important external stimuli impinging on an animal are those emanating from other animals. While everyone feels they know what is meant by communication, it has proved an elusive phenomenon to define. The following may be recognised as key components of a communicatory act.

1 The **sender**: an individual who transmits a signal.
2 The **receiver**: an individual whose probability of behaving in a particular way is altered by the signal.
3 The **signal**: the behaviour (e.g. posture, vocalisation) transmitted by the sender.
4 The **channel**: the medium through which the signal is transmitted (e.g. visual, auditory or electrical).
5 The **context**: the setting in which the signal is transmitted and received (e.g. antagonistic behaviour, courtship).
6 **Noise**: background activity in the channel which is irrelevant to the signal being transmitted.
7 The **code**: the complete set of possible signals and contexts.

6.1 Territorial displays

Animals may communicate with each other in territorial encounters. In some cases the signals may be persistent, so that the individuals concerned need not meet. Scent marking, from specialised glands, for example, occurs in **territorial marking** in many social insects, antelopes and carnivores. In other instances, territorial signals are transient, allowing rapid and alterable exchange of information.

In David Lack's classic book *The life of the robin* (1943), he recounts how he discovered that the display of the robin, *Erithacus rubecula*, functioned not in courtship, as had often previously been reported, but in territorial defence by the owner of a territory towards an intruding robin of either sex. The most prominent feature of the display is the red breast, which is stretched and so held that the intruding robin sees as much of it as possible (Figure 6.1).

In great tits, *Parus major*, territory is proclaimed by male great tits mainly by their songs. Krebs, Ashcroft and Webber on three occasions experimentally removed territorial male great tits from a six hectare piece of mixed deciduous woodland (Higgin's Copse) near Oxford in early spring. One part of the woodland, the control area, was then left alone, one part had a tape-recorder playing one type of great tit song, and the third area had a tape-recorder playing a recording of several great tit songs. In each experiment the control area was the first to be colonised by a new male and the third area with the

Figure 6.1 Threat display in robins. The attacking bird, on the left, is shown higher than, at the same height as, and lower than the other bird.

repertoire of songs, the last to be colonised. Evidently one function of great tit song is to deter territorial intrusion. Krebs suggests a repertoire of songs may be more effective than a single song, due to the *beau geste* effect. In P. C. Wren's adventure novel *Beau geste*, defending troops of the French Foreign Legion propped up their dead comrades at the battlements, thereby convincing the attacking Arabs that they were much more numerous than they actually were. In the same way, if a territory owner successfully mimics the songs of several individuals, he may be able to keep other birds from invading his territory.

Humans can be considered to advertise their territories at three levels: societal, family and personal (Morris, 1978). It is true that by the criteria of exclusive use, defence and the presence of a fixed area (section 4.1), one can argue whether humans are territorial at all. At worst, however, the label 'territory' provides a useful focus for a discussion of certain human behaviour.

First, the societal territory. We evolved as group-living animals, in units of probably only a few dozen individuals at most. As these bands swelled into tribes, and tribes into nations, war chants became bugle calls and national anthems, while territorial boundary-lines hardened into fixed borders. Patriotism, however, is rarely enough to satisfy us. Uncommon is the individual who does not belong to a smaller, more personal sub-group such as the Women's Institute, the Working Men's Club, a teenage clique or a specialist association. Often such groups have territorial signals – badges, headquarters, slogans and all the other displays of group identity.

Second, the family territory. Essentially, the family is the breeding unit. In a typical house the bedrooms are upstairs and therefore relatively safe, distant from the front door where contact is most likely to be made with the outside world. Like other territorial spaces, the family territory has conspicuously displayed boundary-lines such as garden fences and walls. When a visitor crosses these boundary-lines he puts himself in a position where he must ask permission to do simple things, like sitting down, which he would consider a right in a neutral environment. Many families are unable even to visit the seaside or enjoy a picnic without setting up a temporary territory, advertised by rugs and towels, to which they can return from their wanderings.

Third, the personal space. Each of us carries with ourselves, wherever we go, a 'portable' territory called a **personal space**. If people move inside this space, we feel threatened. Jammed into a lift with strangers we relinquish our personal space, but adopt special techniques. We ignore the people invading our territory, minimise the possibility of physical contact, remain silent and adopt an expressionless set of facial features, avoiding eye contact. Different cultures have different ideas about exactly how close one person should stand to another. In England, most people talk at roughly fingertip distance. In Mediterranean countries, the comfortable distance may be only half this. Embarrassment may ensue when, for example, a British diplomat meets an Italian or an Arab diplomat at an embassy function. Many an embassy reception is apparently dotted with threatened western-European diplomats pinned to the wall by rejected Mediterranean diplomats. As with societal and family territories, personal spaces may be advertised even if the territory holder is absent. In one library experiment, placing a pile of journals on a table by a seat successfully reserved the place for an average of 77 minutes. If a jacket was draped over the back of the chair, the 'reservation effect' lasted for over two hours.

6.2 Pre-fight communication

Fights may be dangerous. Communication may serve to ensure that fighting is undertaken, so far as is possible, only when advantageous. In red deer, for example, adult males during the autumn rut may fight for possession of females. Before two stags fight they roar at one another. This is done alternately for several minutes, each male roaring and then listening attentively as his rival roars. Clutton-Brock and Albon investigated the possibility that roaring might be a form of communication. It was suggested that the stronger a male, the greater the rate of roaring he could manage. Evidence for this came from the observation that the most successful males roared the most often, while fights only occurred between males with similar roaring rates. Playback experiments, using tape recorders, showed that males matched, as best they could, the roaring rates they heard. Roaring seems to be a form of **honest assessment**.

After roaring at each other, rival males might continue their antagonism no further, or proceed to a parallel walk (Figure 6.2). Parallel walks are **ritualised**. The two stags move into a tense walk parallel to each other, typically 5–20 metres apart. Their hair is often raised and their gait is slow, regular and stiff.

Figure 6.2 A parallel walk between two adult red deer stags.

Assessment probably continues during parallel walks which can be as short as five seconds or as long as half-an-hour. Only a proportion of parallel walks (about 50 %) leads to fights. Like roaring contests, they are commonest when both contestants hold harems (and therefore have much to lose or gain by fighting). They are also associated with situations where the outcome of the fight is uncertain: they occur more frequently between well-matched opponents than between those that obviously differ in fighting ability, and longer parallel walks are less likely to be followed by fights than shorter ones, but if fights do follow such long parallel walks they are more likely to be protracted.

6.3 The bee language controversy

More than 2000 years ago, Aristotle pondered the phenomenon of honeybee recruitment. He found that although a source of food placed near a hive might remain undiscovered for hours or even days, once a single bee had located the food, many new bees soon appeared. In a series of classic experiments von Frisch investigated how such communication occurred.

Distance communication

When workers collect high-quality food within about 80 metres of the hive, on returning they execute a **round dance** (Figure 6.3). Because it is normally dark in the hive, other workers do not watch the round dance from a distance, but follow it about on the comb as well as sensing vibrations produced by the dancer. The result is that the workers may fly out of the hive to search for the food source. They do not search randomly, in the sense that they only search within about 80 metres of the hive. However, the recruited bees are given no more precise information about the distance to the food source, nor are they

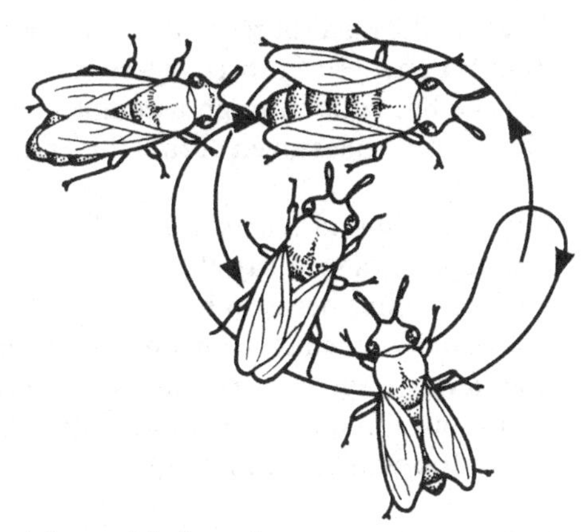

Figure 6.3 The round dance of the honeybee.

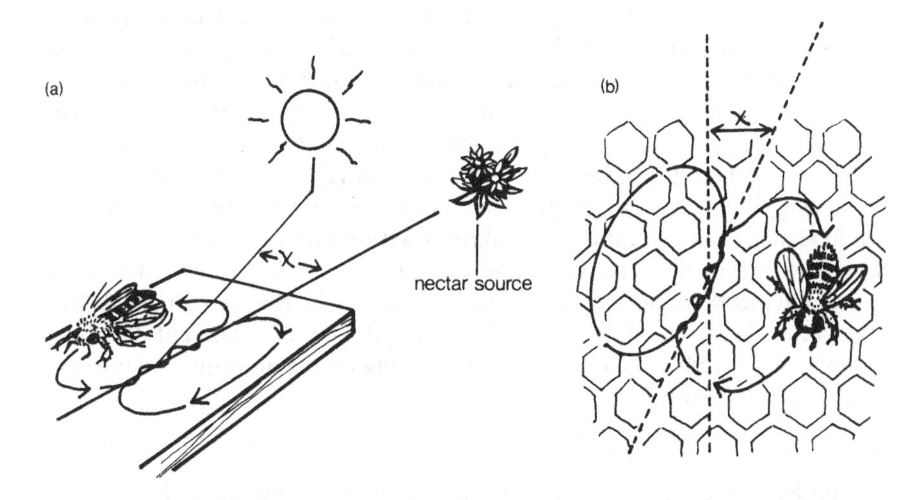

Figure 6.4 The waggle dance of the honeybee. (a) The directional component of the dance is most obvious when the dance is performed outside the hive on a horizontal surface. Here the bee runs directly at the food site. (b) On a comb, inside the dark hive, the dance is oriented with respect to gravity.

given any information about the direction of the food source. Their search is helped by the fact that they pick up odour cues adhering to the body of the dancing bee, and may taste regurgitated nectar.

The round dance does, however, convey information about the profitability of the food source. The higher the quality of the food source, the more often the returning foragers change the direction of their dances. The poorer the food source, the lower the rate of dance reversal.

If a bee finds a rich food source more than about 80 metres from the hive, on its return it performs a **waggle dance** (Figure 6.4). This conveys information about the precise distance to the food source in the range of roughly 80–1000 metres. The information is conveyed in three different codes.

1 The speed at which the bee runs through a complete dance circuit.
2 The number of abdominal waggles given during the straight-run portion of the dance.
3 The frequency with which sound bursts are produced while dancing.

It would, in fact, be more accurate to say that the waggle dance provides information about the energy that needs to be expended on the flight to the food source. If the bee has to fly upwind or up a hill to get to the flowers, the recruiting worker shows fewer waggles, performs its dance circuits in a more leisurely way, and gives a lower frequency of sound bursts than if bees had to fly the same distance with no wind or on level ground.

Direction communication

The waggle dance, but not the round dance, conveys information about the direction of the food source (Figure 6.4). In the dark hive the angle at which the straight-run portion of the waggle dance is performed relative to the vertical equals the angle that the food source makes (clockwise) with the sun. Von Frisch found that bees could even give and receive such information on cloudy days. This is because their eyes are sensitive to polarised light. Polarised light is available on cloudy days, and from it the direction of the sun can be deduced.

It might be thought that one minor problem with the directional component of the waggle dance would be that as the sun moves relative to the position of the food source, a recruiting bee's directional information would gradually become more inaccurate with time as it remained dancing in the hive. Amazingly, workers can correct for this effect, so that if they remain inside the hive occasionally dancing, they perform their dances in a different direction relative to the vertical each time, thus correcting for the movement of the sun relative to the flowers.

Objections to von Frisch's theories

Despite early experimental support for the bee 'language' hypothesis, later experiments carried out by Wenner and his co-workers suggested that recruits locate food source by olfaction alone. Wenner's group did not dispute von Frisch's finding that the waggle dance contains information about the distance and direction of food; but that recruits make use of this information. Wenner pointed to several instances where the behaviour of animals contains information decipherable to humans, but not used by conspecifics of the signaller. For example, the movement of flies having exhausted a food source conveys information to a human observer about the type and concentration of the food source, but other flies do not appear to use this information. Although recruited bees respond to the dances, Wenner suggests that it merely stimulates them to search for food. Once they start searching, food sources are located by olfaction.

Wenner's criticisms stimulated others to attempt new experiments to test von Frisch's hypothesis further. One series of experiments providing emphatic support for von Frisch was performed by Gould. Gould made use of the fact that if a small point source of light is provided at the side of the vertical comb, bees treat this as if it is the sun. This means that when dancing bees treat this point source just as the bees dancing at the nest entrance on a horizontal surface treat the sun (Figure 6.4) and angle their waggle run directly at the food source. Having provided such a light source, Gould then blacked out the three ocelli or simple eyes on the dorsal surface of the head of some workers returning to the hive from rich food sources. Such bees could not detect the point source of light and so oriented their waggles as normal with respect to gravity. The other bees with ocelli, however, followed the dances and interpreted them as if the light source was the sun. In Gould's experiments, every 30 minutes the light on the comb was moved a measured angle. As predicted, the recruited bees' flights were shifted by the same amount.

Despite Gould's experiments, there is still some controversy about the extent to which bees use information from the waggle dance, as criticisms have been levelled at Gould's work and other studies purporting to provide support for the language hypothesis (Rosin, 1984).

6.4 Communication as manipulation

It is easy to start thinking that communication is for the benefit both of the sender and of the receiver. It is important to see this as an oversimplification. Senders will often be selected to deceive (Krebs and Dawkins, 1984). In pre-fight communication, for example, each protagonist should attempt to persuade its opponent that he or she is the stronger. Consider, for example, roaring in red deer (section 6.2 above). Presumably in the past natural selection has acted to increase the roaring rate, even if this has not been accompanied by greater fighting prowess.

Krebs and Dawkins argue that communication can be better understood in terms of selection for effective manipulation rather than in terms of selection for effective information transfer. They suggest that ritualised signals are analogous to human advertising signals. On advertising, transfer of accurate information, apart from the name of the product, is less important than persuasion of the viewer or listener. The features that are found to lead to effective advertising include information redundancy (in which the same information is presented more than once), rhythmic repetition, bright packaging and supernormal stimuli. Such features are also found in a great many animal signals.

Granted the existence of such deception on the part of senders of information, receivers have presumably been selected for their ability to detect and ignore such deceptions. Krebs and Dawkins use the term 'mind-reading' to describe what an animal is doing when it attempts to predict what another animal will do next. Animals should become sensitive to the fine clues by which other animals' behaviour may be predicted. Humphrey has developed the idea of animals as 'nature's psychologists', and goes so far as to suggest that the whole faculty of subjective consciousness and self-awareness evolved as a device to facilitate reading the minds of others.

6.5 Language

At first sight, communication in humans appears fundamentally different from communication in other species. On closer inspection, however, two modifications to this assertion can be made. First, much of the exchange of information between humans is by **non-verbal communication**, and the richness and precision of non-verbal communication in humans is not fundamentally different from that which occurs in other species. Secondly, chimpanzees and gorillas, and possibly some other species, show a capacity to exchange information in a manner which indicates that they have the potential, at least, to communicate in a fashion not wholly distinct from the most abstract of human communication, as discussed below.

Darwin realised that humans have a system of non-verbal communication which plays an essential part in our social lives. It involves postures, gestures and facial expressions of considerable complexity and subtlety. To take one example, depending on its context and the precise way in which it is used, raising one's eyebrows may indicate greeting, suggest approval, invite flirtation, register indignation, express curiosity or admit surprise. Non-verbal communication may or may not be conscious. Both men and women register approval of perceived objects by enlarging their pupils. Accordingly, people prefer pictures of people of the other sex whose pupils have been painted larger than in reality.

Communication using 'true' language has traditionally provided a clear-cut separation of humans from other animals. By true language is meant both the use of **symbols** (e.g. written or spoken words) for abstract ideas, and an understanding of **syntax**, so that symbols convey different messages depending on their relative positions. Numerous investigators have explored the potential of chimpanzees to learn language. Chimpanzees lack the motor ability to produce a sufficient variety of human sounds to enable them to speak more than a handful of words. Different approaches, however, have demonstrated that they do not lack the ability to deal with the other aspects of learning true language. The Gardners used the American sign language for the deaf and their chimp, Washoe, learned over 100 words. Premack taught another chimp, Sarah, to use different plastic blocks as word symbols. It was found that Sarah sometimes intentionally tried to deceive other individuals by her communication. Such lying had previously been thought by some to be a truly human propensity.

A major controversy has arisen over the extent to which chimpanzees and gorillas really can create sentences and indulge in true language. Whatever the outcome of this debate, it is obvious that humans have a richness of symbolic communication orders of magnitude greater than that of any other studied species. Language in humans can truly be said to lie at the heart of our society.

7 Human society

Many studies of animal societies have been carried out within the framework of evolutionary theory. Throughout this book animal behaviour has been discussed from this perspective. Although anthropologists have tried to study primitive tribes within an evolutionary framework, the analysis of developed societies has not generally been tackled in this way. In fact, sociologists view modern society from a number of different perspectives. Their analyses of society depend on the particular perspective they have adopted. The outlines of these perspectives will be discussed briefly later in this chapter.

Human societies have much in common with those of monkeys and apes. All complex social groups require some regulation of relationships within the group. Thus, for example, both human and ape societies recognise kinship and have some kind of dominance system, or at least a form of leadership. However, man differs from monkeys and apes in that he has language, as discussed in the previous chapter, and relies on learned behaviour to a greater extent. Most of man's behaviour, knowledge and beliefs are **cultural**, and are socially rather than genetically transmitted. On the other hand, it is not true that ape and monkey societies have no culture. For example, a group of Japanese macaques, unlike other groups, wash their sweet potatoes in the sea to remove the sand before eating them. This behaviour was first observed in a single, young female. Other members of the group copied her and now infants learn how to wash potatoes from their mothers and the behaviour has become part of the culture of that particular group of monkeys. Man, however, has a much more varied and complex culture than that of primate societies. The possibility for complexity is greatly enhanced by the possession of language. This enables humans to transmit very complex ideas, beliefs and behaviour from one generation to the next.

7.1 Socialisation

In comparison with other species, the human infant spends a much longer period physically and emotionally dependent on his major caregiver, usually his mother. During this time the child must learn about the culture within which he is growing up and how to relate to other people. The process by which an individual learns to become a member of society is called **socialisation**. This is a life-long process. Charles Horton Cooley, in the early part of the century, coined the phrase **primary socialisation** to refer to the learning that takes place, mainly in childhood, within the context of close bonds, such as those with family. He used the phrase **secondary socialisation**, which continues on into adulthood, to refer to the learning which takes place in the context of more formalised relationships such as school or work.

Although this distinction is of some use in theory, to differentiate the contexts within which socialisation occurs, in practice it is not necessarily possible to distinguish two different processes of socialisation, or to say that all individuals learn the same things within the two contexts. What is learned, and when and how it is learned, may vary from individual to individual. Childhood socialisation will be discussed in greater detail in the next chapter.

7.2 Norms, roles and status

In order to analyse the structure of society, sociologists and anthropologists have developed a number of **concepts** to refer to aspects of society. Concepts are abstract ideas but nevertheless do relate to what goes on in the real world. Clearly defined concepts are useful tools in that they enable sociologists to refer simply to complex ideas. For example, socialisation, described above, is a concept. Since they are clearly defined, concepts may also help in describing society and developing useful theories and generalisations. However, their use may force sociologists to view a society in a particular way, which may not necessarily be appropriate. It is, therefore, important to be aware that the use of particular concepts may lead to unhelpful assumptions about society. For example, the use of the terms primary and secondary socialisation may be misleading if it promotes the assumption that the two are separate, distinct processes.

Obviously, different societies have very different cultures. However, the pattern of social existence in most societies can usefully be described using certain key concepts. All societies have a set of **norms**. These are patterns of behaviour common in the society, which are unwritten and usually accepted without question by the members of that society. Norms differ widely from one society to another. The norms of one society may seem very peculiar to another. For example, the eating of dead relatives practised by some primitive tribes seems very macabre to us. The most important norms in a society, which are usually concerned with the way members of the society behave towards each other and involve moral values, are called **mores** – for example, the belief that we should not kill. Less important norms (such as, in our society, shaking hands) are called **folkways**. A description of the norms, mores and folkways of a society is helpful in trying to understand that society. This kind of description is often a very important part of an anthropologist's study of a society.

All individuals in society are involved in numerous different relationships with others. Sociologists use the concept of **role** when exploring these relationships. A role is the expected behaviour associated with an individual who occupies a particular **status**. Status may either signify a social position, such as child, woman or maried person, or it may designate a position of esteem or prestige, such as prime minister, headmistress or chairman of the board. A status may either be **ascribed**, if honour is accorded to someone as a result of birthright, such as a king, or **achieved**, if an individual gains esteem by his own merit, for instance as an eminent scientist. An individual plays a number of different roles during his lifetime, some consecutively. For example, a woman starts life playing the role of a girl. This changes as she grows up. Once an adult, she may be a wife, a mother and, in her job, a teacher. Just as an actor interprets a part and thus may play his role in a different way from another actor, despite reading from the same script, so one individual may play a slightly different role to another, even when they hold the same status. Sociologists describe this by saying that despite some accepted **role demands** in society, individuals define roles in their own way, which affects their **role performance**.

The concept of role is useful in trying to understand the structure of relationships within society. A role can rarely be played by an individual on his own. The nature of how the role is played is partly determined by the behaviour of other people towards the actor. So, for example, the role performance of a teacher is partly dependent on the behaviour of her pupils, colleagues, parents and headteacher. This kind of cluster of related roles is known as a **role set**. By breaking down the relationships within society into role sets it may be easier to start to understand how these relationships work.

Sometimes an individual may find himself in a **role conflict**. This occurs when behaviour expected of him or her in the context of one status clashes with that expected in another. One common conflict is that between the roles of wife and mother. For example, as a wife a woman may be expected to cook dinner for her husband but at the same time she is needed to look after her child. The resolution of role conflicts can be important in the continuing development of an individual. For instance, a teenage girl may be torn between becoming a wife and mother at an early age and going on to higher education. The existence of role conflicts in society can also be important in the understanding of, for example, change in society. For instance, the development of feminism has been affected by the consistent conflicts between the roles of wife, mother and working woman.

Thus, the general behaviour of individuals in a society may be described in terms of norms, mores and folkways. The structure of relationships may be depicted in terms of interconnected statuses and their corresponding roles. The development of the individual and that of society is affected by the changing status and connected roles experienced during the individual's lifetime. A significant part of adult socialisation is the learning of new roles. Most adults will be faced with learning at least two new roles during their lifetime: becoming a husband or wife and changing their job.

Successful marriage usually requires a degree of mutual interdependence and mutual compromise between husband and wife. On marriage, both husband and wife will not only have to learn what is required of them in those roles but also, to some extent, to adjust their attitudes and behaviour to those of their spouse to promote a harmonious relationship. For example, the wife may prefer her husband not to go out most nights with his male friends and the husband may prefer his wife not to be involved in sport every weekend, so that the two can spend time together. Adult socialisation can be thought of as resocialisation, as it may require a modification of behaviour learned during childhood. Lack of adjustment and poor communication between spouses is an important source of marriage breakdown and so their successful accomplishment is obviously important.

A change in job may also require considerable resocialisation. This is not just in terms of learning what is expected in the job but also how to fit in with work-mates or colleagues and what sort of behaviour is expected, peripheral to the job, but at the place of work. For example, it is a joke among teachers that one of the most important things a new teacher has to learn is who sits in which seats in the staff room. Resocialisation may be particularly necessary for someone who not only changes job but also changes status, for example moves from managerial to manual work. Of course, occupational change may also involve a move to another part of the country, which in turn may require further adjustment. As these examples show, socialisation or resocialisation are constant processes that continue 'from the cradle to the grave'.

7.3 Sociological perspectives

As has been mentioned earlier, sociologists describe society in different ways depending on their particular perspective. These perspectives are not necessarily contradictory but simply place the emphasis on different aspects of society. It is, therefore, helpful when trying to understand society to look at it from the point of view of all the perspectives. It is beyond the scope of this book to discuss the perspectives in great detail but it is important to have some understanding of the general principles behind the most common perspectives.

There are three main sociological perspectives. These are (a) the functionalist perspective; (b) the conflict perspective; and (c) the interactionist perspective. The **functionalist perspective** was first described by Emile Durkheim. He likened society to a living organism. The various **social institutions**, such as the family, education and religion, function together for the benefit of the whole society, in the same way as the organs of the body work together. Order and balance are regarded as normal to society. There is general agreement or **consensus** among members of society about behaviour, values, norms, roles, etc. Change is thought to occur in society only when it is necessary for society to function smoothly. Change is evolutionary, not revolutionary. Functionalists argue that an individual's behaviour is determined by his experiences within social institutions, such as the family, school and work. Functionalist sociologists aim to analyse and explain how social institutions function and contribute to the smooth working of society.

The functionalist perspective emphasises consensus. In contrast, as implied by its name, the conflict perspective emphasises **conflict** as the driving force of society. The structure of society is seen as resulting from the resolution of past conflicts between its members. It is regarded as constantly changing as one group presses for change and another opposes it. In other words, change in society is thought to be revolutionary. The most important analysis of the conflict perspective has been made by Karl Marx. He argued that all but the most primitive societies are made up of two **social classes**. One is a ruling class which exploits the other. The ruling class owns the **means of production**, in other words the land, factories, raw materials, etc. which are needed to produce the goods necessary for the smooth running of society. The ruling class dominates the economy while the exploited class simply works for the ruling class. Change in society occurs as a result of the class struggle between the two groups. The Marxist argues that the individual's source of identity is rooted in his class membership. Conflict sociologists aim to analyse, describe and explain class conflict.

The interactionist perspective differs from the previous two in that it emphasises the importance of the actions and interactions of individuals within society. The first seeds of this perspective were sown by Max Weber in the 1920s. Rather than seeing people as wholly influenced by society, interactionists believe that individuals have a degree of freedom of action and that people's feelings and behaviour as individuals can influence the way society develops. They argue that it is important to find out what people think about society and how they interpret others' actions, as this can provide useful information about how society works. Interactionist sociologists are not such a unified group as functionalist or conflict sociologists; nevertheless, they all agree about the importance of studying the individual rather than groups.

It is unlikely that the three perspectives will be reconciled into one viewpoint. They present three different faces of society, all of which are valid. A full picture of human society will probably require an analysis from each point of view.

7.4 Social stratification

As far as is known, all societies, past and present, are divided into social layers, producing inequalities among members. This is known as **social stratification**. These divisions are usually based on economic factors, such as wealth and income, prestige in the form of honour and status, and power – though gender and race may also divide society. All the sociological perspectives agree that social stratification exists but they disagree as to its significance to society and how it should be studied. Functionalists believe that social inequality is necessary and useful to society. They see the relationship between different strata as being one of interdependence and co-operation. On the other hand, conflict theorists believe that the different layers are essentially in conflict and that social struggle is inevitable. The interactionist is less concerned with the relationship between the strata but more with how individuals see their positions within the groups and the groups' relationships with each other.

7.5 Hindu caste system

There have been, and are, many different forms of social stratification. One of the earliest was the system of slavery, as was common in, for example, ancient Greece and Rome in which there were two strata: the slaves and their owners. The feudal system of medieval Europe is another example of a somewhat more complex system. One highly complex, rigid system is the **Hindu caste system**. Some aspects of this system still influence life in India today but much of it has broken down. Members of society were divided according to occupation into five main strata or castes.

1 Brahmins: the priesthood.
2 Kshatriyas: rulers and administrators.
3 Vaishyas: traders, merchants and farmers.
4 Shudras: servants and manual workers.
5 Untouchables: social outcasts doing very menial jobs.

Each of these castes were then further subdivided into sub-castes or **jati**, which were groups involved in particular occupations. A man was born into a jati and this was the only way of acquiring membership. Membership of caste was therefore ascribed. Movement between strata was extremely rare. It could only occur when a woman married into a higher caste. Societies where social stratification is rigidly fixed at birth are often referred to as **closed societies**.

The caste system is supported by the Hindu religion, which teaches that children are born into a particular caste because that is where they deserve to be and that contact with a lower caste results in harmful pollution. Nevertheless, the caste system is essentially organised round occupation, like almost all other types of social stratification. This system is worth studying because it is something of an archetype of stratification systems. It is very rigid and there is a tendency for the particular sub-castes to have developed their own cultures, which renders them even more separate from each other. The caste system has also proved to be extremely resilient to attempts to break it down.

7.6 Social class

Modern industrial societies, for example Britain, are divided up into social classes. These societies are sometimes referred to as **open societies** since, in comparison to systems such as the caste system, position is determined to a greater extent by achievement and there is more opportunity for movement between strata. Social classes are often considered to be mainly economic groupings but the way of life and social relationships within them are also different. Thus, it is vital to study social class in order to understand not only social organisation but also individual social relationships.

The first major analysis of social class was given by Karl Marx. Marx believed that the social classes are defined in terms of their relation to the means of production (section 7.3). In a capitalist society the ruling class, which generally owns the means of production, is called the **bourgeoisie** and the subject class the **proletariat**. Marx argued that class struggle was inevitable and that in due course the proletariat would overthrow the bourgeoisie. However, some people have suggested that over time class divisions have become less pronounced and that as members of the proletariat become more wealthy, they are being absorbed into the bourgeoisie. Thus revolution is becoming less likely. This theory of **embourgeoisement** will be discussed in section 7.8 below.

In contrast to Marx, who saw class as determined purely by economic factors, Max Weber argued that class is three-dimensional. He suggested that society is divided in three ways: by class, status and party. Class is determined by economic factors and is dependent on the individual's access to scarce goods and services. Status is determined by social position and is dependent on prestige and esteem. Lastly, party is a political grouping created by the individual's power. An individual may attain power by a number of different means, for example, as a politician or as a result of his or her position in industry. Weber argued that in order to understand divisions within society all three forms of stratification, which are not necessarily independent, should be considered. Despite their differences, both Marx and Weber thought that the roots of stratification can be explained in terms of specific factors. They also agreed that class affects many other areas of life.

Because, as has been stressed, the influence of class spreads into most aspects of life, it is very difficult to come up with a simple way of dividing society. The most commonly used method is by occupation. This is reflected in our day-to-day assessment of other people. One of the first questions we ask about someone is 'What does he do?' Once we know the answer to this, we immediately make predictions about a whole range of aspects of that person's life, for example his education, attitudes and family background. Of course, we make these predictions from the stereotyped view we have of each class. Nevertheless, there is some basis in fact for most assumed class differences. However, even if those differences did not exist in reality, the very fact that most people assume they exist may reflect the way they think about and react to others, depending on where in the social structure they perceive them to be.

Sociologists also use occupation as the main basis for defining class. A

Table 7.1 Registrar General's fivefold classification of social class

Class	*Type of occupations*	*Percentages*		
		1951	1961	1971
1	*Professional and higher administrative* e.g. lawyers, architects, doctors, managers, university teachers	3	4	4
2	*Intermediate professionals and administrative* e.g. shopkeepers, farmers, actors, musicians, teachers	18	15	18
3	*Skilled* (a) Non-manual (1971: 21.1 %) e.g. draughtsmen, shop assistants, clerks (b) Manual (1971: 28.4 %) e.g. electricians, coalminers	50	51	49
4	*Semi-skilled* e.g. milk roundsmen, bus conductors, telephone operators, fishermen, farm workers	17	21	21
5	*Unskilled* e.g. nightwatchmen, porters, refuse collectors, cleaners, labourers	12	9	8

number of different scales have been devised. The one most commonly used is that of the Registrar General. This uses a fivefold classification but divides class three into two groups: manual and non-manual workers (see Table 7.1). The original Registrar General's scale was devised in 1911 and there have been changes since then. Some new occupations have been added and existing ones moved up or down the scale. However, it is often very difficult to assess exactly where a particular occupation should be placed. For example, should postmen be placed in Class 3 (skilled) or Class 4 (semi-skilled) categories? Another well-known scale is that developed by John Hall and D. Carog Jones in the 1950s. This is a seven-point scale (see Table 7.2).

Obviously, the greater the number of class divisions made in a scale the more possibility there is for fine adjustments. On the other hand, in a more complex scale it can become extremely difficult to decide in which class a particular occupation should be placed. This may then mean that particular scales are biased in favour of particular groups, depending on the predictions of the author. For example, some sociologists consider that the Hall–Jones scale is biased in favour of non-manual workers. There are four classes for them and all are placed above manual workers. In comparison, the Registrar General's scale places both manual and non-manual workers in Class 3. Some biases are understandable, in that they reflect the groups of society that a particular sociologist is most interested in studying. Nevertheless, it is important to realise that although most people are agreed that class divisions exist, to tie all

Table 7.2 Hall and Jones' sevenfold classification of social class

Class	*Type of occupation*	*Example*
1	Professional and high administrative	Company director of large firm
2	Managerial and executive	Civil servant
3	Inspectional, supervisory and other non-manual, higher grade	Teacher
4	Inspectional, supervisory and other non-manual, lower grade	Insurance agent
5	Skilled manual and routine grades of non-manual	Clerk
6	Semi-skilled manual	Agricultural worker
7	Unskilled manual	Barman

occupations down into precise categories, satisfactory to everyone, is not necessarily possible.

Occupation is most commonly used as the basis for dividing society, and many differences in lifestyle between the classes stem from differences in the workplace. In general non-manual workers enjoy better working conditions, better fringe benefits, longer holidays and greater chance of promotion compared with manual workers. The middle classes also tend to have higher pay, less supervision and greater job security. The effects of these differences spill over into home life. For example, lack of job security makes it more difficult to get a mortgage and to plan for the future; lack of an adequate pension scheme may make life for the retired manual worker very different from his non-manual counterpart; manual workers, who have poorer work conditions and less contact with their superiors compared with many of the middle class, may be more motivated to join a trade union.

Closely linked to occupation is, of course, income. Marked differences between classes in terms of income do occur; for example, currently in Great Britain, those in social class 1 (professional and higher administrative workers) are likely to receive, on average, 3.5 times the earnings of those in social class 5 (unskilled workers). However, there is sufficient variation to make amount of income probably one of the least satisfactory indicators of class. For example, nowadays a skilled manual worker may earn considerably more than a schoolteacher. Neither is the distribution of wealth (i.e. assets that can be bought and sold) a good way of assessing class. This is because the distribution of wealth is so unequal that it will only distinguish a very small group at the top of society. The Royal Commission on the Distribution of Income and Wealth (The Diamond Commission, 1974–79) found that the top 1 % of the population owns a quarter of all the wealth in Great Britain. The bottom 80 % of the population owns less than a quarter. Contrary to some opinion, wealth has not been redistributed to any significant degree in recent years. Much power is still in the hands of the wealthy few, but because they are only a few, wealth is not much help in trying to distinguish the class of the vast majority of people.

Where redistribution of wealth has occurred, this is as a result of the vast increase in home ownership in recent years. Nevertheless, home ownership, and access to housing in general, still distinguishes the classes. In 1980, more than 80 % of professional and managerial workers were owner-occupiers, whereas only 26 % of unskilled workers were in that position. At the very bottom of the scale millions of people, particularly those of ethnic minorities, live in substandard housing. Many find it difficult to get council accommodation.

For those who still say that class divisions in Britain have largely disappeared, one telling distinction certainly still exists. The class a person belongs to influences that person's chances of staying alive. Life expectancy, infant mortality and health are closely linked to class. In all cases the working classes fare worse than the middle classes. Table 7.3 shows the deaths of men by major cause and social class between 1970 and 1972 in England and Wales. Only in the cases of suicide and obscure cancers do the middle classes have a greater than average risk and then only slightly. The incidence of disease is

Table 7.3 Chart to show deaths of men by major cause and social class in England and Wales between 1970 and 1972. Standard mortality ratios, i.e. actual deaths against expected numbers randomly distributed, are used

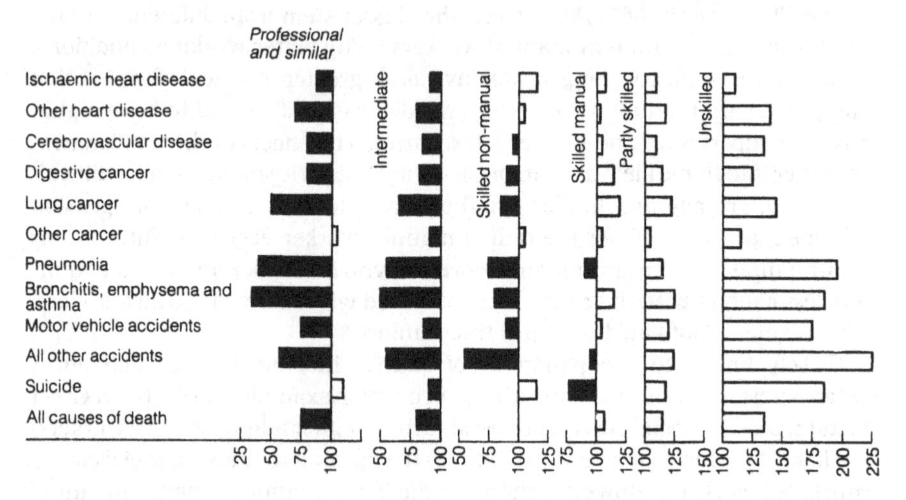

related to housing and diet. The working classes tend to have both poorer housing, as has already been discussed, and poorer diet, perhaps as a result of lower incomes. The middle classes also tend to make better use of National Health Service care available. Despite the levelling of the classes in some other aspects, the gap for life expectancy has widened in recent times and is still widening.

Our society supposedly provides equality of opportunity in education but nevertheless this is another area where class differences are apparent. The middle classes both make better use of the system and succeed better within it. For example, in 1977 36 % of students accepted for universities through UCCA came from the professional classes, whereas only 24 % came from families where the father was a manual worker. The reasons for this discrepancy are numerous; only a few can be mentioned here.

Differences in family lifestyle and attitudes between the classes can affect how children see school, their motivations to succeed within it and the practical help they get from school to succeed. For example, parental encouragement, presence or absence of books at home, attitudes to authority, time, space and quiet to work can all affect a child's success. Schools in middle-class areas often have better facilities, attract better teachers and offer an environment more conducive to learning than schools in working-class areas. At the top end of the scale, private, fee-paying schools offer an education highly geared to gaining their pupils' success in the academic sphere. Within schools, it has been argued that many teachers are biased towards middle-class children: lessons may be more appropriate for them and teachers may constantly expect them to do better than working-class children, which in itself can promote middle-class success and working-class failure. Thus, teachers' opinions can become 'self-fulfilling prophecies'.

In our society, success at school and the passing of exams is almost a prerequisite to occupational success particularly in the professions, for example, medicine or the law. A child who fails at school is unlikely to get a good job. Thus, inequalities at school can help perpetuate class divisions. However, it must be remembered that a good education system is also very important for breaking down divisions and encouraging mobility between classes. This will be discussed in section 7.7 below.

In addition to easily measurable differences between classes mentioned above, it can also be argued that there are cultural differences between the classes. This is particularly evident when comparing the working class and the middle class. The classes differ in terms of accents and use of language, taste (for example, the way in which they decorate their houses), manners and etiquette, and in attitudes. For example, traditional working-class groups tend to have a strong neighbourhood and family life, promoting a sense of solidarity and an 'us' and 'them' attitude to life, where 'them' refers to the upper classes. Child-rearing techniques may also vary. For instance, working-class parents tend to use and threaten physical punishment more than middle-class parents and to be less likely to use reason and persuasion and to give explanations for rules. This may sometimes lead to a mistrust of figures of authority among working-class children. As has been mentioned, this can lead to difficulties for these children at school. However, it is important to be aware that these kinds of differences are essentially differences in class stereotypes and are not applicable to everyone in their entirety. In fact, traditional working-class culture is disappearing in Britain today. It should also be noted that the very existence of class stereotypes is divisive in itself. If people expect to find differences, they may exaggerate them and reject someone from another class out of hand.

7.7 Social mobility

Social mobility is the movement of individuals between social groups or classes. This happens to some extent in an open society such as Britain or the USA. It may be assessed in a number of different ways. **Intergenerational mobility** measures an individual's movement to a higher or lower social class from that of his father. **Intragenerational mobility** measures an individual's movement between classes during his lifetime. **Stratum mobility** measures the movement of an occupational group from a higher or lower position in the stratification system. The first two types of social mobility focus on individuals whereas the third one concerns the change in status of a whole group, though of course this then affects the individuals within that group.

It is often thought that people in Britain have plenty of opportunity to move up the social scale. However, the findings on social mobility in Britain do not entirely bear this out. The most recent extensive survey of social mobility, the Nuffield Social Mobility Survey, carried out by Dr John Goldthorpe, was published in 1980. This is based on information collected from interviews of 10 000 men aged between 20 and 64. A summary of the findings is set out in Table 7.4. Goldthorpe found that, although more people are reaching the top of the scale now than at any time for 60 years, this is mainly because of changes in the occupational structure or in level of income, not because an individual has a better chance of moving up the social scale as a result of his personal achievement. Goldthorpe found that those born into the bottom two classes were three times less likely, and those born into the middle three classes twice as unlikely, to end up in the top two classes as those whose fathers were members of those top two classes (see Table 7.4). Nevertheless, despite these rather depressing statistics, there is social mobility, both up and down, in British society and the potential for movement is greater than it has ever been.

As has been mentioned, the most important reason for this is changes in the

Table 7.4 Social mobility – distribution of the sample according to subjects' and subjects' fathers' status and category

Sons' class in 1972

Father's class	1	2	3	4	5	6	7	total
1	45.7 25.3	19.1 12.4	11.6 9.6	6.8 6.7	4.9 3.2	5.4 2.0	6.5 2.4	100 0 (680)
2	29.4 13.1	23.3 12.2	12.1 8.0	6.0 4.8	9.7 5.2	10.8 3.1	8.6 2.5	100.0 (547)
3	18.6 10.4	15.9 10.4	13.0 10.8	7.4 7.4	13.0 8.7	15.7 5.7	16.4 6.0	100.0 (687)
4	14.0 10.1	14.4 12.2	9.1 9.8	21.1 27.2	9.9 8.6	15.1 7.1	16.3 7.7	100.0 (886)
5	14.4 12.5	13.7 14.0	10.2 13.2	7.7 12.1	15.9 16.6	21.4 12.2	16.8 9.6	100.0 (1072)
6	7.8 16.4	8.8 21.7	8.4 26.1	6.4 24.0	12.4 31.1	30.6 41.8	25.6 35.2	100.0 (2577)
7	7.1 12.1	8.5 17.1	8.8 22.6	5.7 17.8	12.9 26.7	24.8 28.0	32.2 36.6	100.0 (2126)
Total	100.0 (1230)	100.0 (1050)	100.0 (827)	100.0 (687)	100.0 (1026)	100.0 (1883)	100.0 (1872)	(8575)

Classes:

1 Higher professionals, higher-grade administrators, managers in large industrial concerns and large proprietors.
2 Lower professionals, higher-grade technicians, lower-grade administrators, managers in small businesses and supervisors of non-manual employees.
3 Routine non-manual – mainly clerical and sales personnel.
4 Small proprietors and self-employed artisans.
5 Lower-grade technicians and supervisors of manual workers.
6 Skilled manual workers.
7 Semi-skilled and unskilled manual workers.

occupational structure. In recent years the number of jobs in industry has declined and that in the service sector increased. Increased mechanisation and greater bureaucracy has led to an increase in the number of white-collar, middle class jobs. As a result of **differential fertility** (in other words middle-class families having a lower birth rate than working-class families) the middle class was not able to fill the rapidly expanding number of jobs and so left the way open for successful working-class people to take them. However, since the mid-1960s birth rates have fallen generally and so differential fertility has declined in importance as a boost to social mobility.

Another factor which should affect social mobility is 'equality of educational opportunity'. However, research – for example A. H. Halsey (1980) – has shown that, although access to higher education for everyone has increased over the last 40 years, inequalities still exist. Working-class children do not do as well as middle-class ones. As has been discussed earlier, a working-class background may prove a disadvantage for children at school. A good educational system can potentially be an enormous boost for social mobility but in practice it does not always work that way.

Three other factors affect social mobility to a lesser degree than change in occupational structure, education and differential fertility. These are intelligence or talent, marriage and migration to the cities. Not surprisingly, the possession of high intelligence or a particular talent will help an individual to move up the social scale. On the other hand, a person's social background may often mean that an individual's potential is not fully realised. If it is assumed that the distribution of intelligence is the same for the working class as the middle class, then it is certainly the case that many intelligent individuals with working-class backgrounds are not reaching the same level as those from the middle class with equal talent. Upward mobility through marriage, for example the successful businessman marrying his secretary, is largely confined to women. It is still the case in Britain that many women depend on men for economic support. Thus, more men than women are in a position to marry, for example, for looks, and not for financial reasons. Good looks may confer high status on a woman which can outweigh any lower class background. The third factor affecting social mobility is migration to the cities. Increasing urbanisation as a result of industrialisation is accompanied by an influx of largely unskilled labour into the cities. The people whose jobs they replace are then able to move up into more skilled jobs.

The discussion of social mobility so far has emphasised upward mobility. However, much mobility is downward. As can be seen from Table 7.4, in all classes less than 50 % of sons remain in the same class as their father, though overall there is more upward than downward mobility, as a consequence of changing occupational structure. It should be noted that measuring social mobility is not as simple as it seems. A major problem is deciding at what point in a person's lifetime to assess his social class. Intragenerational mobility can be significant. Most research has simply picked an arbitrary age at which to assess class, which of course could be misleading. Stratum mobility may lead to a son doing the same job as his father but being judged to be in a different class. This, too, may not always be properly represented.

7.8 Embourgeoisement – are we all middle class now?

A number of sociologists have claimed that many working-class people are adopting middle-class attitudes and lifestyles and that hence Britain is becoming increasingly middle class. This argument is based on a number of social, economic and political changes that have taken place in Britain in recent years, some of which have already been discussed. For many, incomes have risen in real terms and, therefore, so have living standards. Acute poverty is now more rare. Power in industry is now vested less in owners than in managers and many companies are publicly owned. Power in general is dispersed over a larger number of people. Class conflict, as predicted by Marx, has waned. Increased opportunities in education have led to a greater possibility of reaching a high social position through merit.

However, it can be argued that most of these changes have not been radical enough to promote embourgeoisement on a large scale. In the 1960s John H. Goldthorpe, David Lockwood et al. conducted a large study of industrial workers in Luton in order to test the embourgeoisement theory. They came to the conclusion that despite the relatively high wages of these workers, they were not becoming middle class. There are three important reasons for their conclusions. First, unlike many of the middle class, these workers did not consider their jobs to provide any satisfaction in themselves but simply to be a source of income. Second, they had not in general adopted middle-class lifestyles. Third, there was no evidence that they had changed their voting pattern. Most still voted Labour rather than Conservative and saw Labour as their class party. It must also be remembered that to change class an individual does not just have to change his perspective but also has to be accepted by members of the class to which he aspires. This is often the most difficult thing of all and can present a barrier to all social mobility. Nevertheless, despite scepticism about the embourgeoisement theory, it is fairly clear that there is at least some convergence of classes in Britain today. The result, however, may not necessarily be a class that completely resembles the present middle class.

8 The individual in society

In the previous chapter the structure of society, particularly in Britain, was discussed. In this chapter some aspects of the development of individuals within society will be discussed. As they grow up, children learn both to be the same and to be different from others. They learn to be the same in that they learn about the norms, values and common roles in society but they also learn to be different in that they develop their own distinctive personalities. Family, school and peers all contribute to these two sides of socialisation.

Psychologists no longer see the child as a passive object upon which these **agents of socialisation** act but as an active individual who also affects the behaviour of those around him. Thus, the child influences those around him which in turn affects the way in which those around the child behave towards him. Right from the moment he is born the baby engages in interaction with others and it is through this interaction that he learns about society and develops his personality. However, a baby is born with predispositions to learn certain things. He may also come into the world with a tendency towards certain temperamental characteristics, such as level of activity or irritability. To some extent, parents' reactions to their children will depend on these characteristics, and this starts babies off on their individual paths of development.

This book concentrates on examining the structure of society and the role of the individual within it. Therefore, this chapter will mainly discuss the development of personality within the context of society and, to some extent, the influence of personality on relationships. Only a brief mention will be made of the theories of personality and the controversies surrounding them.

8.1 Attachment

Most infants' first close social contacts are with the mother. Therefore, for the rest of the chapter, the mother will usually be assumed to be the primary caregiver. Whether it is essential that the mother is the primary caregiver has in fact been a matter for much debate. Many psychologists believe that the type of relationship and nature of **attachment** that develops between a mother and her baby in the early months affects the child's later relationships and the development of his personality. By attachment is meant a relatively enduring emotional tie to a specific other person.

A baby of a few days old can recognise his mother by smell, and a little later also distinguish her voice. At the age of two months the baby will smile happily at the sight of his mother's face. However, a baby of this age is apparently equally delighted by any face. It is not until he is about four months that he begins to prefer familiar to unfamiliar faces. By six or seven months he clearly knows those familiar to him and will show distress when separated from them. Between the ages of seven and twelve months an infant becomes increasingly focused on specific people to whom he has become attached. The attachment figure, usually the mother, provides comfort and a secure base from which the child can explore. During the first half of the second year, protest over mother leaving is at its height. After this, the infant gradually gains more independence and becomes less distressed at a separation from his mother.

At one time the most important factor in attachment was thought to be feeding. It was assumed that an infant becomes attached to the person who regularly feeds him. However, it now appears that this is not necessarily the case, as a set of experiments with rhesus monkeys, carried out by Harlow and Harlow in the 1960s, suggests. The Harlows separated infant monkeys from their natural mothers at an early age and raised them with artificial surrogate mothers. The surrogates were dummies made to look something like real monkeys. Some of them were simply made of wire mesh and others were padded with foam rubber and covered with soft terry cloth (Figure 8.1). The monkeys showed a clear preference for the soft padded mother even if they

Figure 8.1 An infant monkey's response to an artificial mother. Although fed via the wire mother, the infant spends more time with the terry-cloth mother. The terry-cloth mother provides security and a safe base from which to explore.

were fed from a bottle attached to the wire mesh mother. Given the choice, the monkeys would always go to the padded mother when distressed. Those monkeys raised only with a wire mesh mother showed more disturbance, rocking and clutching themselves and failing to explore, than those reared with the padded mother. Clearly a warm, comfortable area to which to cling was more likely to promote attachment than a source of food.

A number of psychologists have used a research procedure known as the **strange situation** to study the way in which infants are attached to their mothers. This was first devised by Mary Ainsworth in the 1960s. In this a mother and her infant are brought into a strange room, well supplied with toys. After a few minutes an unfamiliar person enters and engages the mother in conversation. The stranger then tries to play with the child and a little time later the mother leaves the room. The child's reaction to the stranger while the mother is gone, and his response to his mother on her return, are noted. Hundreds of infants have now been observed in this situation and from their behaviour three types of attachment relationship have been identified.

Type A. **Avoidant**. Before separation the infant pays relatively little attention to his mother. He is not particularly distressed when his mother leaves and either ignores her or greets her very tentatively on her return.

Type B. **Securely attached**. Before separation the infant plays quite happily and reacts positively to the stranger. After the mother leaves the infant plays much less and is obviously distressed. On the mother's return the infant goes to her quickly, is calmed and returns to play.

Type C. **Resistant**. Here the infant is fussy and wary before the mother leaves. On her return the infant may go to her quickly but then resist contact and show anger, struggling when picked up or hitting the mother. The infant will not then resume play easily.

Ainsworth and others have found that Type B is the most common, though there is some variation between cultures. Mothers may contribute to the development of particular attachment relationships by the degree of sensitivity and responsiveness they show to their infants. Securely attached infants usually have mothers who are the most responsive.

The issue of how far the nature of early attachment affects later development is a controversial one. Clearly, maternal deprivation or long-term separation from the mother has devastating effects. The infant monkeys reared by Harlow and Harlow with surrogate mothers showed clear disturbance in their later social behaviour. When placed with normally reared age-mates these monkeys either withdrew in fright, or were unusually aggressive, or both in turns. They were unable to engage in normal sexual activity, appearing not to know how to behave. The males never bred successfully. The few females who managed to breed proved to be extremely incompetent mothers. However, even limited contact with other infants during the period of separation from mothers appeared to alleviate the worst of the developmental consequences. Monkeys who spent some time with age-mates, although still withdrawn and frightened to explore, were more at ease with other monkeys and were able to mate successfully.

More limited separation from the mother also appears to have some long-term consequences. Hinde and Spencer-Booth studied the effects of short-term separation in rhesus monkeys. Infants were separated from their mothers for one or two six-day periods at about seven months of age. These monkeys were observed with their mothers at 12, 18 and 30 months, and their behaviour compared with a control group of monkeys who had not been separated from their mothers. At 12 months the previously separated monkeys stayed closer to their mothers in the home cage than the controls, but the differences were small. However, in a test in an unfamiliar cage differences between the groups were much more apparent. The mother and infant were placed in an unfamiliar cage that had a tunnel big enough for the infant but not the mother to go through. The previously separated infants were much more wary of going down the tunnel than the controls. By 30 months most differences between the monkeys had disappeared, but previously separated monkeys were still more nervous at approaching an experimenter and less prepared to try to get food out of reach, when on their own, than the controls. This study shows that, although long-term effects of separation may not show up under normal living conditions, previously separated animals may be less able to cope when faced with a stressful situation, such as being on their own or facing a strange individual.

The studies by Harlow and Harlow show, not surprisingly, that extreme forms of deprivation of maternal contact have severe consequences. The studies of short-term separation demonstrate that a very mild form of isolation may also have some long-term consequences. However, neither is definite evidence that a failure to form an attachment or a disruption of an existing attachment may impair a human infant's social functioning for life. Some psychologists, in particular John Bowlby, have argued that a failure to form a close attachment, preferably to the mother, early in life may result in an inability to develop close relationships in adulthood. Bowlby also argued that continuity of the attachment relationship is very important. He suggested that children who suffer disruptions to this relationship may then experience sudden depressions and anxieties in adulthood that are unrelated to events in their lives at this later date. Bowlby's views have been enormously influential since they were adopted by legal systems worldwide as the basis for making decisions on the placement of children who come to their attention. For example, a belief in the importance of maintaining continuity in the attachment relationship may lead to children being allowed to remain with their natural parents even if they have been abused. More recent research suggests that although separation is very distressing for children over the age of six months, if new, satisfactory attachments can be formed after separation then no long-term effects can be detected.

8.2 Child-rearing practices

Before making any comments on the effects of child-rearing practices on the developing child's personality, it is important to point out that there is not necessarily a 'right' and 'wrong' way to bring up children, nor are there clear 'good' and 'bad' personalities. A child's behaviour should be appropriate to the culture and circumstances in which he will have to live. Even within the same culture some people prefer different personalities to others. Studies of child-rearing practices suggest generalisations about the kind of children that tend to be associated with particular parental behaviour. This does not necessarily mean that all children from a certain kind of family will turn out in the same way. Children are influenced by many other aspects of their environment other than their parents, for example school and friends. It is not easy to sort out the web of factors that may affect children and so be able to make successful predictions about their later personalities. This chapter will simply give a flavour of some of the research findings about child-rearing practices that have come out in recent years.

In the 1960s and 1970s, Diana Baumrind and her colleagues, on the west coast of America, carried out a large-scale study of children and their parents. These nursery school age children were observed with their parents at home and in the laboratory. They were also observed at their nursery school. As a result of these observations at school the children were rated on a number of characteristics, such as 'impetuous', 'self-reliant' or 'happy'. Parents were also rated on the way in which they behaved towards their children. In her later studies Baumrind distinguished three patterns of parenting. These are:

1 **Authoritarian.** Parents who fit into this category are likely to try to control their children with little discussion. They believe in obedience and an absolute set of standards.
2 **Authoritative.** Parents in this category attempt to control their children by reason. They recognise that children as well as adults have rights. They believe in independence and self-direction in their children. They do not necessarily see themselves as infallible but do not base their decisions mainly on the child's desires.
3 **Permissive.** Parents in this category are likely to make few demands of their children and use little punishment. They accept their children's impulses and behaviour and try to avoid controlling them.

Children of these types of parents were found to differ in a number of ways. However, there were some sex differences, in other words boys could respond in a different way to a particular parental behaviour than girls, and vice versa. In general, though, authoritarian parents had children who showed little independence and were average on social responsibility scores (e.g. helping, sympathy). Authoritative parents had children who were independent and socially responsible. The children of permissive parents were low on social responsibility and not particularly independent.

Some of the children who were studied at nursery school were then followed up at age eight or nine. Again, these children were observed at school and while with their parents. At this age children were rated according to what Baumrind called social and cognitive agency. A child was rated high in social agency if he or she showed evidence of leadership and was bold and not anxious in his or her interactions with schoolmates. A child was rated high in cognitive agency if he or she responded well to intellectual problems, was oriented towards achievement, had a sense of identity and showed evidence of original thought. The relation between parental behaviour and children's social and cognitive agency is shown in Table 8.1. Generally speaking, children who are high in agency do not usually have permissive parents. As before, there were sex differences. Girls were more likely to score highly on the social and cognitive agency rating if they often argued with their parents. Boys were more likely than girls to lose interest in achievement and social relationships if their parents had been authoritarian.

The studies by Baumrind's group are only a small number amid a large research area. The findings discussed emphasise differences in children associated with different parental strategies for control. However, there are obviously many other aspects of parental behaviour that may affect a child's development. Two other areas, which have also been stressed, are communication and warmth. Parents have widely differing styles of communication. Some, such as many of the authoritative parents in Baumrind's studies, have very open communication with their children, where things are discussed and the children allowed to voice disagreements. Communication style may vary between social classes (section 7.6). Most parents show some affection for their children but the degree of warmth can vary considerably. Research suggests that children who have particularly warm, loving parents are likely to be more altruistic and considerate of

Table 8.1 Relations between parental behaviour and children's social and cognitive agency

Parental practices	*Children's characteristics at ages 8–9*	
	Boys	Girls
At preschool age:		
Authoritative	Fairly high cognitive and social agency	Very high cognitive and social agency
Authoritarian	Mid-level social, low cognitive agency	Mid-level cognitive and social agency
Permissive	Low social, very low cognitive agency	Low cognitive and social agency
At ages 8–9:		
Firm rule enforcement		High social and cognitive agency
High demand for self-control	High cognitive agency	
Authoritarian	Low social and cognitive agency	

schoolmates at an earlier age and have higher **self esteem**, i.e. feelings of self worth, than those from less loving families. A child's need for love and affection is very strong. Thus, withdrawal of love is a very powerful tool that parents have for controlling their children. In the short term, withdrawing love, such as refusing cuddles when they are wanted, can produce compliant children. However, in the long term, it can produce anxiety in the children and may lead to less co-operation.

Obviously, to some extent, parents differ in their child-rearing practices because of differences in their own personalities and attitudes. However, there are also some more general factors that may influence how parents treat their children. Parents' social class has already been mentioned as a factor particularly affecting communication and control. Previous experience as parents may also influence how a parent behaves. There is evidence to suggest that first-borns are treated rather differently from subsequent children. For example, at the same age, first-borns may be talked to and disciplined more often. nunger children may be treated as babies for longer than first-borns. The age of the child in general will also affect parental behaviour. Obviously, parents do not treat their ten-year-olds in the same way as they treat their one-year-olds. A parent's view of what to expect from children of particular ages will affect how their behaviour changes, how they interpret their children's behaviour and the sort of allowances they make for incompetent or naughty behaviour. A fifteen-month-old who tips his dinner all over the floor will usually be treated very differently from an eight-year-old who does the same thing. Parents' expectations for appropriate behaviour may well influence when that behaviour occurs in their children. Expectations may also affect the development of the child's **self concept**, which will be discussed in section 8.4 below.

Before ending this brief discussion on the influence of parents on their children, it is important to note that children also influence their parents. The socialisation process is a two-way one. Parents may be more or less responsive to their children but equally children may be more or less responsive to their parents. A child's behaviour will inevitably affect his parents' actions. For example, highly aggressive children tend to receive high levels of punishment from their parents. Research suggests that this is not just because parental punishment may promote aggression but also because the children are aggressive for some other reason and this encourages punishment in the parents.

8.3 Peers and siblings

In the early months of life, the primary influence on the infant is that of his parents. From the second year on, the influence of **peers** and siblings becomes increasingly important. Until recently research has tended to concentrate on the influence of parents. However, there is now a rapidly expanding interest in the role played by peers and siblings in development. As in the previous section, this section will simply give the reader a taste of the relevant ideas.

A child's relationship with his parents is obviously very important to his development but his relationship with others, particularly peers and siblings, may be equally so. They may, in fact, contribute a different dimension to development compared with parents. The child's relationship with adults, particularly his parents, can be seen as a complementary one. The parent is dominant and nurturing whereas the child is dependent. On the other hand, the child's relationship with peers is one between equals. This difference between peers and parents has been emphasised by a number of developmental psychologists, in particular Jean Piaget. He saw this difference as explaining why children often learn different things from their parents compared with their peers.

Adults tend to instruct the child in the rules of social life, such as customs and others' expectations of him. There is usually little negotiation between children and their parents. However, with peers, a child comes to learn that he can actually create a social world, invent rules, change meanings and negotiate with others. He comes to realise that society is not necessarily immutable but can be changed and influenced by individuals such as himself. He is better able to learn what it is like to have power. This is illustrated by the fact that children involved in games are often more concerned with devising and changing the rules than with actually playing the game (Figure 8.2). As one psychologist, John Youniss, put it, 'Children come to cope with and understand the social world either *through* adults or *with* peers'. A child's relationship with his siblings is in a sense a special case of his relationship with peers. Although he is in many ways on an equal footing with his siblings, his relationship with them is likely to be much more intense. He has been with his siblings all, or most of his life, and so is likely to know them much better than any of his peers.

Figure 8.2 Children at play.

Most babies first come across others of the same age during their first year. However, at this stage other infants are mostly ignored or treated as if they were inanimate objects. Smiling and touching does occur but these are very intermittent. One-year-olds are more likely to interact with other children but there is little evidence to suggest that children of this age actively seek others to play with. However, between the ages of two and five children make enormous strides in their capacity for social relationships. Interactions with peers become sustained and complex. Children of this age also show the beginnings of friendship with others rather than just indiscriminate playing with anyone, although at this age friends are usually described purely in terms of what they do, such as 'They play with me' or 'They live next door'.

Once at school children's concepts of friendship rapidly develop and they incorporate notions of mutual interdependence and co-operation into their definition of friendship. By adolescence peers, particularly friends, have probably become more socially important than parents or other adults. Typically, clashes occur between teenagers and their parents because their friends do one thing and their parents want them to do another. At this stage, role conflict for the adolescent, between being a son or daughter and being a friend, is common. Throughout childhood peers are an important influence in the socialisation process but in adolescence they are perhaps most significant.

8.4 Self concept

One important aspect of personality upon which peers and parents may have a different influence is the self concept. Our self concept is essentially our view of ourselves. That view may be affected by anything from our material possessions to our attitudes. It includes self esteem, which is our evaluation of ourselves, of our abilities and worth relative to other people. Self concept also includes self recognition, usually shown by an ability to recognise ourselves in a mirror, and a sense of personal identity. This is our sense of continuity over time. It also involves a notion of **ideal self**, the kind of person we would like to be.

Most theorists argue that a child's self concept develops through his relationships with others. Parents may give their children basic aspects of their self concept, such as recognition of whether they are a boy or a girl, and they may also promote high or low self esteem. However, it is through peers that the child develops a more realistic notion of himself. The importance of social relationships for the development of the self concept was particularly stressed by George Herbert Mead, writing in the 1930s. He argued that without being part of society a person could not develop a self concept. He believed that through social interaction the child begins to understand that others have a view of him, for example that he is 'good' or 'naughty'. The child may take over these views and incorporate them in his view of himself. Once he is able to do this, he may then be able to think about the views others have of him: he is then looking at himself as if he were a third party. Mead considered that once the child reached this stage he had truly developed a notion of self. This ability to see oneself as others see one, and to adjust behaviour accordingly, is essential for the smooth running of social relationships. Without this capability we would constantly be involved in conflict and misunderstanding. Later, in addition to looking through the eyes of particular individuals, a child comes to see how he measures up to the norms of the 'generalised other' or society to which he belongs. He will then **internalise**, or take over as part of himself, the norms and expectations of society. Thus, as was mentioned at the beginning of this chapter, society promotes similarities and differences in children.

More recent research on the development of the self concept in children has suggested that children go through several recognisable stages during this development. The first stage is the acquisition of self recognition. From about the age of 18 months children are able to recognise themselves in a mirror. Next, they come to understand that their own thoughts are private and not shared by others. Very young children can be frustrated because they assume that everyone must know what they are thinking. Once they start talking, children begin to define themselves in terms of their characteristics. Initially, these are purely physical or external, for example their sex, what they look like and where they live. From the age of about six or seven they begin to include psychological characteristics in their lists, for example if they are shy or happy. Gradually, children's self concepts become differentiated, in that they come to recognise that they may be good at some things or in certain respects and bad at others. They learn to adjust their behaviour according to others' expectations of them. Finally, children develop an ideal self, the kind of person they would like to be.

It is difficult to disentangle how parents and peers may influence the development of the self concept but research has given some pointers. Studies have shown that children with high self esteem tend to have parents who are accepting and approving, who can be strict but allow the children to express their opinions. Parents with children who have high self esteem also tend to use reasoning or withdrawal of privileges rather than physical punishment or withdrawal of love as a means of discipline. However, children with high self esteem also tend to have a history of success in what they have done. It is, therefore, difficult to know whether parents treat these children as they do because of the way they behave or because, genuinely, that type of parental behaviour results in high self esteem for their children. The most likely explanation is that both are true to some extent. It is clear that children who suffer rejection by other children tend to have low self esteem and a relatively poorer view of themselves than more popular children. However, as yet there is little evidence to explain exactly how peers influence the development of personality and the self concept for the vast majority of children. There are only a number of general theories, such as that of Mead, described above, as to how it might be done.

8.5 Personality and relationships

Psychologists have always studied personality. However, there has always been, and still is, controversy about how to describe it. In the early days people were classified into types, for example in terms of the kind of body build they had, which was considered to be associated with personality. More recently

trait theory has become more popular. Traits are characteristics in which one individual is assumed to differ from another in a relatively permanent and consistent way, such as aggressiveness or emotional stability. A number of psychologists, for example Hans Eysenck, have used a statistical technique, known as factor analysis, to reduce the number of possible characteristics. Eysenck, for example, considered that there are two basic dimensions of personality: **introversion–extroversion**, and **stable–unstable**. The characteristics associated with these two dimensions can be seen in Figure 8.3. Characteristics such as these are assessed by means of **personality inventories**. These are questionnaires containing a series of questions, the answers to which are related to particular characteristics.

Despite years of work on constructing inventories designed to measure traits, they are in fact very poor predictors of behaviour. Correlations between test scores and measures of behaviour or others' assessment of personality rarely rise above 0.3. There are a number of possible explanations for this. One is that personality traits are not in fact stable but that a person behaves differently depending on the situation. A number of psychologists consider that the situation is more important in determining behaviour than individual personality traits. They argue that consistency would be found, and therefore correlations would be higher, if only similar situations were observed. For those psychologists who do believe in consistent traits, there is the possibility that the same traits are not important for everyone. If one is studying a number of individuals, all of whom are consistent on different traits, then across all those individuals, measured on the same traits, correlations between traits and behaviour would be low. However, there is evidence that much higher correlations can be obtained if only those traits that each individual considers to be significant and consistent for himself are considered. In other words, different traits are used for each individual.

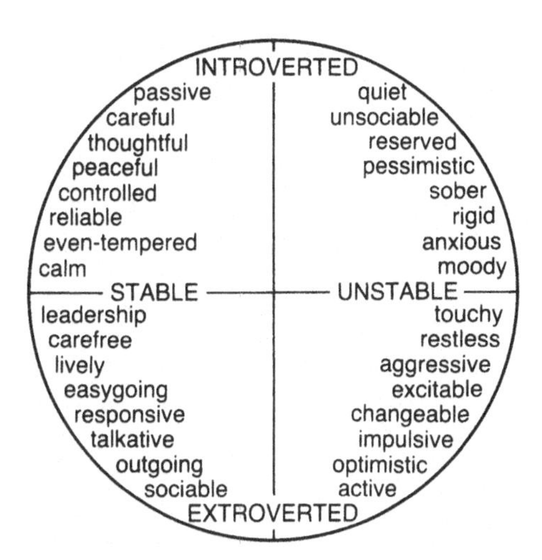

Figure 8.3 Eysenk's dimensions related to personality traits. Various traits are shown in relation to the two basic dimensions of introversion–extroversion and stability–instability.

Even from this short discussion it can be seen that defining, measuring and studying the adult personality is difficult and few clearcut findings have emerged. If it is not even clear how to investigate the adult personality, it is of course even more difficult to chart the development of personality. Nevertheless, it is important that research should be done in this field. The kind of people we are, is bound up with the kind of relationships we have with others. This, in turn, affects the kind of society that we live in. Relationships are then the prime source of childhood socialisation. Thus, the study of personality and relationships is important both for understanding society and for understanding individuals.

9 Epilogue

This book has covered a wide range of social behaviour in a large number of different species of animals, including man, and has discussed ideas and research from a wide variety of, often separate, disciplines. It is important to try to link the ideas, research and disciplines and not to create divisions and distinctions that may hinder understanding.

As was discussed at the beginning of the book, a particular example of an inappropriate distinction that has often been made is that between instinct and learning. It is sometimes helpful to distinguish behaviour which appears more or less fully formed the first time an animal finds itself in the appropriate situation from that which clearly needs to be modified through learning. However, even the most rigid behaviour may be influenced by the environment, either before or after the behaviour first appears. Instinct and learning are not like numbers in an addition sum that can be filled in to give an explanation for particular behaviour. In the same way the 'biological' and 'cultural' aspects of human social behaviour and society cannot be specifically quantified. Biological factors can only be seen as providing the predispositions or foundations for behaviour. Whether it is the biological, sociological or psychological factors that are investigated will depend on the interests of the investigators, the particular behaviour they are studying and the type of question they wish to answer. In the end, all kinds of explanations are likely to contribute to our understanding.

Throughout the book the behaviour of man and other animals has been discussed side by side. In a number of cases, such as aggression and sexual behaviour, knowledge of, and explanations for, the behaviour of animals has been used to try to understand man's behaviour in apparently similar situations. This comparative approach may sometimes be helpful. A study of animals is particularly useful where an experimental manipulation of behaviour can be done that cannot be carried out with people. However, man is very different from other animals in that he possesses language and an enormous capacity for learning. A direct comparison with other animals should, therefore, only be made with caution. This book has directly and indirectly made such comparisons but the reader must judge for himself or herself whether they are really valid in all cases.

Further reading and references

Atkinson, R. L., Atkinson, R. C. and Hilgard, E. R. (1983) *Introduction to psychology* (8th ed), Harcourt Brace Jovanovich (Section 4)

Bateson, P. (ed) (1983) *Mate choice*, Cambridge University Press

Bertram, B. (1978) *Pride of lions*, J. M. Dent

Bilton, T., Bonnett, K., Jones, P., Sheard, K., Stanworth, M. and Webster, A. (1981) *Introducing sociology*, Macmillan (Chapters 2 and 12)

Booth, T. (1975) *Growing up in society*, Methuen

Braithwaite, A. and Rogers, D. (1985) *Children growing up*, Open University Press (Sections I and IV)

Clutton-Brock, T. H., Guinness, F. E. and Albon, S. D. (1982) *Red deer: the behavior and ecology of two sexes*, Chicago University Press

Darwin, C. (1859) *The origin of species by means of natural selection or the preservation of favoured races in the struggle for life*, John Murray

Dawkins, R. (1976) *The selfish gene*, Oxford University Press

Fossey, D. (1983) *Gorillas in the mist*, Penguin Books

Halliday, T. R. and Slater, P. J. B. (eds) (1983) *Animal behaviour, vol. 3: genes, development and learning*, Blackwell Scientific Publications

Hinde, R. A. (1982) *Ethology: its nature and relations with other sciences*, Fontana Paperbacks

Jarvis, J. U. M. (1981) 'Eusociality in a mammal: cooperative breeding in naked mole-rat colonies', *Science* 212, 571–3

Krebs, J. R. and Dawkins, R. (1984) 'Animal signals: mind-reading and manipulation'. In Krebs, J. R. and Davies, N. B. (eds) *Behavioural ecology* (2nd edn), Blackwell Scientific Publications, 380–402

Lack, D. (1943) *The life of the robin*, H. F. and G. Witherby Ltd

Maccoby, E. E. (1980) *Social development: psychological growth and the parent-child relationship*, Harcourt Brace Jovanovich

McFarland, D. (ed) (1981) *The Oxford companion to animal behaviour*, Oxford University Press

Manning, A. (1979) *An introduction to animal behaviour* (3rd edn), Edward Arnold

Martin, P. and Bateson, P. (1986) *Measuring behaviour. An introductory guide*, Cambridge University Press

Morris, D. (1978) *Manwatching: a field guide to human behaviour*, Triad Granada

Nobbs, J. (1983) *Sociology in context*, Macmillan Education

O'Donnell, M. (1983) *New Introductory Reader in Sociology*, Nelson Harrap (Section C)

Popplestone, G. (1985) *Social issues in British society*, Heinemann

Reiss, M. J. (1984) 'Courtship and reproduction in the three-spined stickleback', *Journal of Biological Education* 18, 197–200
Slater, P. J. B. (1985) *An introduction to ethology*, Cambridge University Press
Sylva, K. and Lund, I. (1982) *Child development: a first course*, Blackwell
Tinbergen, N. (1974) *Curious naturalists*, Penguin Education
van Lawick-Goodall, J. (1971) *In the shadow of man*, Collins
von Frisch, K. (1967) *The dance language and orientation of bees*, Harvard University Press
Watts, C. R. and Stokes, A. W. (1971) 'The social order of turkeys', *Scientific American* 224(6), 112–8
Wilson, E. O. (1975) *Sociobiology: the new synthesis*, Harvard University Press
Wilson, E. O. (1978) *On human nature*, Harvard University Press
Woolfenden, G. E. and Fitzpatrick, J. W. (1984) *The Florida scrub jay: demography of a cooperative-breeding bird*, Princeton University Press

Index

INDEX